To order or receive additional information on these or any other
McGraw-Hill titles, in the United States please call 1-800-822-8158.
In other countries, conta Waterford RTC *e.* **BC15XXA**

3 9003 00058788 0

Multimedia Network Integration and Management

Multimedia Network Integration and Management

Larry L. Ball, Ph.D.
Managing Director
Telenetic Controls, Ltd.
Vancouver, B.C., Canada

McGraw-Hill

New York   San Francisco   Washington, D.C.   Auckland   Bogotá
Caracas   Lisbon   London   Madrid   Mexico City   Milan
Montreal   New Delhi   San Juan   Singapore
Sydney   Tokyo   Toronto

Library of Congress Cataloging-in-Publication Data

Ball, Larry Lennox.
    Multimedia network integration and management / Larry L. Ball.
        p.   cm.— (McGraw-Hill series on computer communications)
    Includes index.
    ISBN 0-07-005227-1
    1. Multimedia systems.   2. Computer networks.   I. Title.
II. Series.
    QA76.575.B35   1996
    006.6—dc20   95-46467
                                                                    CIP

McGraw-Hill

A Division of The McGraw-Hill Companies

1 2 3 4 5 6 7 8 9 0   AGM/AGM   9 0 1 0 9 8 7 6

ISBN 0-07-005227-1

*The sponsoring editor for this book was Marjorie Spencer, the editing supervi-
sor was Stephen M. Smith, and the production supervisor was Suzanne W. B.
Rapcavage.*

Printed and bound by Quebecor/Martinsburg.

To my mother, Virgina:
Thanks, Mom

Contents

Preface   XV

Chapter 1. Multimedia Systems Introduction   1

1.1 .... Definitions: Multimedia, Hypertext, Hypermedia   2
1.1.1 .... Multimedia   2
1.1.2 .... Hypertext and Hypermedia   9
1.1.3 .... Personal Multimedia Systems   12
1.1.4 .... Networks and Integration   14
1.2 .... Overview of Essential Technologies   14
1.2.1 .... Enabling Technologies   14
1.2.2 .... Implementation Technologies   19
1.2.3 .... Multimedia Database Management Technologies   20
1.2.4 .... Data/Image Compression Technologies   21
1.2.5 .... User Platform Technologies   22
1.2.6 .... Transmission Systems Technologies   23
1.3 .... Classifications and Standards   24
1.3.1 .... Hierarchy of Multimedia Creations   24
1.3.2 .... Categorization of Media Effects   26
1.3.3 .... Multimedia Data Standards   26
1.3.4 .... Organizational Forms of Multimedia Data   29
1.4 .... Issues Involved in Multimedia Network Management   30
1.4.1 .... Psychophysical Aspects of Multimedia   30
1.4.2 .... Vendors Involved   30
1.4.3 .... Operational Restrictions   31
1.4.4 .... Multimedia Quality   31
1.4.5 .... Multimedia Network Management Paradigm   34
1.5 .... System Considerations for Multimedia Implementations   35
1.5.1 .... Hardware   35
1.5.2 .... Multimedia Communications Concepts   37
1.6 .... A Look Ahead   39

Chapter 2. Multimedia Technologies:
Audio, Text, and Images · 45

2.1	The Technologies of Multimedia Systems	46
2.2	On-Line Storage Technologies	50
2.2.1	Compact Disk Technology	50
2.2.2	Magnetic Storage Technology	51
2.3	Image Collection Technologies	51
2.3.1	Document Scanners	51
2.3.2	Optical Character Recognition	53
2.3.3	Frame Grabbing	53
2.3.4	FAX Capability	54
2.4	Image Processing	54
2.4.1	Image Enhancement Using Histogram Modification Techniques	57
2.4.2	Image Enhancement Using Image Smoothing Techniques	57
2.4.3	Image Enhancement Using Low-Pass Filtering Techniques	61
2.4.4	Image Enhancement Using Signal Averaging Techniques	62
2.5	Animation Facilities	64
2.6	Other Technologies	66
2.6.1	Knowledge-Based Systems	66
2.6.2	Theorem Proving	66
2.6.3	Automatic Programming	66
2.6.4	Solutions to Combinational and Scheduling Problems	67
2.6.5	Machine Perception	67
2.6.6	Expert Consulting Systems	68
2.6.7	Robotics	68
2.6.8	Natural Language Processing	68
2.6.9	Computer-Aided Instruction	69
2.6.10	Topological Reference	69
2.6.11	Text-to-Speech Conversion	69
2.6.12	Voice Recognition	69
2.6.13	Tone Decoders	70
2.7	Basic Components of Multimedia Services	70

Chapter 3. Multimedia Technologies -
Video Systems · 75

3.1	Evolution and Importance of Video	76
3.2	Basic Video Parameters	79
3.2.1	Video Generation	80
3.2.2	Video Presentation	81
3.3	Emerging Video Services	84
3.3.1	Video Dial Tone	84
3.3.2	Packet Video and Audio Accounting Systems	86

3.4. .... Video Storage Technologies   86
3.4.1 .... Laser Disk Technology   86
3.4.2 .... WORM Devices   87
3.4.3 .... CD-ROM Technology   87
3.4.4 .... Optical Disks   87
3.4.5 .... Magnetic Disk Storage Technology   88
3.4.6 .... Video Jukeboxes   88
3.5. .... Video and Image Collection   89
3.5.1 .... Video Cameras   89
3.5.2 .... Document Scanning   90
3.6. .... Video Processing   90
3.6.1 .... Fourier Transforms   92
3.6.2 .... Neural Networks   92
3.7. .... Video Transmission Technologies   95
3.7.1 .... Fiber Optics   95

Chapter 4. Multimedia Technologies: Multimedia Data Management   99

4.1 .... On-Line Storage Technologies   100
4.1.1 .... Physical Organization   100
4.1.2 .... Traditional Data Storage and Retrieval Schemes   102
4.1.3 .... Video Information File Structuring   107
4.2. .... Attributes of Multimedia Database Systems   108
4.2.1 .... Speed   108
4.2.2 .... Capacity   110
4.2.3 .... Multimedia Information Buffering   111
4.2.4 .... Ease of Use   112
4.2.5 .... Processing Environments   113
4.3. .... Features of a Multimedia Data Management System   114
4.3.1 .... Database Systems Interfaces   115
4.3.2 .... Complexity and Compatibility   116
4.4. .... Example of an Image Database in Operation   116

Chapter 5. Multimedia Technologies: Transmission Systems   121

5.1 .... Transmission Systems for Multimedia   122
5.1.1 .... ADSL/HDSL   122
5.1.2 .... ATM/SONET   122
5.1.3 .... Coaxial Cable   125
5.1.4 .... Twisted-Pair Copper Wire   125
5.1.5 .... Fiber Distributed Data Interface (FDDI)   127
5.1.6 .... Frame Relay   128

5.2..................Attributes of Transmission Systems   129
5.2.1...............Demographic Considerations   129
5.2.2...............Historical Considerations   129
5.2.3...............Technical Considerations   130
5.2.4...............Economic Considerations   131
5.3..................In-Depth Analyses of Transmission Technologies   132
5.3.1...............High-Bit-Rate Digital Subscriber Line (HDSL)   132
5.3.2...............Asymmetrical Digital Subscriber Line (ADSL)   136
5.3.3...............Analog Transmission Systems (ATS)   138

Chapter 6. Multimedia Technologies: Interfacing with Network Systems   143

6.1 ....Existing Network Systems for Multimedia   144
6.2..................AT&T's Unified Management Architecture (UNMA)   144
6.2.1...............Background   144
6.2.2...............Multimedia Technology Impacts   147
6.3..................IBM's Netview   149
6.3.1...............Background   149
6.3.2...............Multimedia Technology Impacts   151
6.4..................GTE's Network Performance Monitor   153
6.4.1...............Background   153
6.4.2...............Multimedia Technology Impacts   155
6.5..................Novell's Netware   157
6.5.1...............Background   157
6.5.2...............Multimedia Technology Impacts   159

Chapter 7. Multimedia Technologies: Coding and Compression   165

7.1 ....Data Conversion   166
7.2..................Pulse Code Modulation (PCM) Techniques   167
7.2.1...............Adaptive Pulse Code Modulation (APCM)   168
7.2.2...............Differential Pulse Code Modulation (DPCM)   169
7.3..................Information Coding   170
7.3.1...............Loss versus Lossless Algorithms   170
7.3.2...............Interframe Coding   171
7.3.3...............Motion Picture Experts Group Coding   174
7.3.4...............Linear Predictive Coding (LPC)   175
7.3.5...............Run-Length Coding   182
7.3.6...............Huffman Coding   183
7.4..................Complexities of Information Coding Algorithms   185
7.4.1...............Delays   185
7.4.2...............Complexity   185

7.4.3 .... Sensitivities to Algorithms and Channel Errors   185
7.5 .... Data Compression   186
7.5.1 .... Null Compression   187
7.5.2 .... Bit Mapping   187
7.5.3 .... Run Length   188
7.5.4 .... Diatomic Encoding   191
7.5.5 .... Pattern Substitution   192
7.5.6 .... Relative Encoding   192
7.5.7 .... Digital Facsimile   193
7.6 .... Digital Audio Compression   194
7.7 .... Frame Compression Techniques   195
7.7.1 .... Interframe Compression   195
7.7.2 .... Intraframe Compression   196
7.7.3 .... Companding (Compression and Expansion of Data)   196

Chapter 8. Applications Requirements for Multimedia Networks   201

8.1 .... General Requirements   202
8.2 .... Components of Multimedia Networks   203
8.2.1 .... Multimedia Network Elements   204
8.2.2 .... Multimedia Network Protocols   206
8.2.3 .... Applications Packages   208
8.2.4 .... Network Management Implications   209
8.2.5 .... Multimedia Information Loading   216
8.3 .... Multimedia Data   217
8.3.1 .... Imagery   217
8.3.2 .... Text   218
8.3.3 .... Audio   219
8.4 .... Issues Related to Multimedia Applications   220
8.4.1 .... Multimedia Data Transmission   220
8.4.2 .... Multimedia Data Reception   220
8.4.3 .... Quality Determination   222

Chapter 9. Multimedia Network Control Techniques   227

9.1 .... Network Control Criteria   228
9.1.1 .... Quality Tagging   229
9.1.2 .... Quality Requirements   230
9.2 .... Error Reporting and Evaluation   232
9.2.1 .... Video   234
9.2.1.1 .... Frame Size   236
9.2.1.2 .... Frame Depth   236
9.2.1.3 .... Compression Ratio   237

9.2.1.4............Noise Level   237
9.2.2...............Audio   237
9.2.2.1...........Musical Instrument Digital Interface (MIDI)   237
9.2.2.2 ....Audio Quality Measures   239
9.2.3..............Text Information   240
9.2.3.1...........Background   240
9.2.3.2 ....Text Quality Measures   241
9.3................Predictive Methods   244
9.3.1.............Polynomial Regression   244
9.3.2.............Correlation Functions   245
9.3.3.............Simulation   246
9.3.3.1...........Single Servicing Points   246
9.3.3.2 ....Multiple Servicing Points   248
9.4................Intelligent Approaches   251
9.4.1.............Data Structures   252
9.4.2.............AI Tools   256
9.4.3.............Knowledge Transfer and Reasoning   257
9.5................Design Control Approaches   260
9.5.1.............Reliability   260
9.5.2.............Availability   262
9.5.3.............Delay Times   263
9.5.4.............Capacity   264
9.5.5.............Execution Speed   266

Chapter 10. Multimedia Network Management Architectures

271

10.1 ....Architectural Concepts   272
10.1.1 ....Open Systems Interconnect (OSI)   272
10.1.2.............The ISO Environment   274
10.1.3.............Common Management Information Protocol   275
10.1.4.............Layer Management Entities   275
10.1.5.............System Management Application Entity (SMAE)   275
10.1.6.............Development Approaches to Multimedia Networks   276
10.1.7.............OSI Model Communications   280
10.2...............Multimedia Network Management   282
10.2.1 ....Multimedia Network Management Requirements   282
10.2.2............Approaches to Multimedia Network Management   285
10.2.3............Component Technologies of Multimedia Networks   286
10.2.4............Protocols for Multimedia Network Management   290
10.2.4.1.........OSI-Defined Protocols   291
10.2.4.2 ....Network Signaling Protocols   294
10.2.4.3 ....Packet-Oriented Protocols   295
10.2.4.4 ....Message-Oriented Protocols   295
10.2.4.5 ....X.25 Data Communications Protocol   297
10.2.4.6 ....TCP Data Communications Protocol   297
10.3...............Access to Network Management Information   299

10.4...............Network Management Schemes 301
10.4.1............ISO Network Management Functions 303
10.4.2 ....Simplified Network Management Protocol (SNMP) 303
10.4.2.1 ....SNMP Description 303
10.4.2.2........SNMP Architecture 304
10.4.3 ....CMIP versus SNMP 305
10.4.4 ....Manufacturing Automation Protocol (MAP) 306
10.4.5 ....Systems Network Architecture 306
10.5...............Operations of Protocols 308
10.6...............Economics of Protocols 309

Chapter 11. Multimedia Network Architectures 315

11.1 ....The Special Nature of Multimedia Networks 316
11.1.1 ....Classification of Generic Networks 317
11.2...............Architectural Configurations 320
11.2.1 ....Architectural Problem Areas 320
11.2.2............Architectural Diversities and Requirements 321
11.2.3............Specific Design Options 323
11.3...............Funcitons of Telecommunications Protocols 327
11.3.1 ....Multimedia Network Issues 330
11.3.2............Network Principles 331
11.3.2.1.........Connectivity 331
11.3.2.2 ....Performance Management 332

Chapter 12. Multimedia Network Scenarios 337

12.1 ....Promise of the Future 338
12.2...............Multimedia Banking Terminal 343
12.2.1............Application Overview 343
12.2.1.1..........Market Opportunities 344
12.2.1.2 ....Key Features and Capabilities 345
12.2.1.3 ....Market View 345
12.2.1.4 ....Network Services 347
12.2.1.5.........Functional Alternatives 347
12.2.1.6 ....Technical Issues 348
12.2.2 ....Network Management Issues 348
12.3...............The Personal Information Booth 350
12.3.1............Application Overview 350
12.3.1.1..........Market Opportunities 352
12.3.1.2 ....Key Features and Capabilities 353
12.3.1.3 ....Market View 353
12.3.1.4 ....Functional Alternatives 354
12.3.1.5.........Technical Issues 354

12.3.1.6.........Development Issues   355
12.3.2............Network Management Issues   357
12.4...............Movies-on-Demand Paradigm   358
12.4.1............Application Overview   360
12.4.2............Business and Revenue Opportunities   364
12.4.2.1.........Analog Copper Delivery System   364
12.4.2.2 ....Digital Copper Delivery System   365
12.4.2.3 ....Analog or Digital Coax Delivery System   365
12.4.3............Network Management Issues   365
12.5...............Personal Video   366
12.5.1............Application Overview   366
12.5.1.1.........Background   366
12.5.1.2.........Customer Need or Problem Addressed   367
12.5.1.3.........Key Features and Capabilities   367
12.5.1.4.........Benefits   368
12.5.2............Market View   368
12.5.2.1.........Description of the Target Market and Potential
                 Customers   368
12.5.2.2 ....Current and Projected Target Market Window and Size   369
12.5.3............Conceptual Approach   369
12.5.3.1.........Provisioning Approaches   369
12.5.4............Key Issues   371
12.5.4.1.........Possible Regulatory/Legal Constraints   371
12.5.4.2 ....Technical Efforts Required   371
12.5.4.3 ....Special/Unique Client Capabilities Needed   372
12.6...............Personal Network Directory   372
12.6.1............Application Overview   372
12.6.2............The Market   373
12.6.2.1.........Historical Perspective   373
12.6.2.2 ....Recent Events   374
12.6.2.3 ....Problem/Opportunity   374
12.6.2.4 ....Product Potential   376
12.6.3............The Product   377
12.6.3.1.........Consumer End Product   377
12.6.3.2 ....System Architecture   378
12.6.3.3 ....User Interface   378
12.6.3.4 ....Operational Scenario   379
12.6.4............The Technology   379
12.6.4.1.........System Design   379
12.6.4.2 ....Key Technologies   380
12.6.5............The Issues   385
12.6.6............Product Definition   386
12.6.7............Service and Support   386
12.6.8............Architecture   387
12.6.9............The Market   391
12.6.10..........Risks and Justification   392
12.6.11 ....Description of the Personal Directory   393

Index   397

PREFACE

The subject of multimedia is like the proverbial unicorn; that is, you hear a lot about it, but you hardly ever see one. The interest and relevance of multimedia to today's telecommunications network management problems is, first, that multimedia functions are not as concerned with the first 5 or 6 layers of the Open Systems Interconnect (OSI) model as is the typical data communications system. Multimedia communications are a blend of interactive request and presentation qualities as well as transmission and/or storage qualities of the system. As a result, multimedia networks, and their coincident management, focus upon the higher layers, such as the applications layer, i.e., layer 7 of the OSI model. Second, multimedia networks lend themselves to restoration when frames of information are received in error, partly because the area of correct and incorrect responses is much less a matter of black and white than is the case with data networks. Third, multimedia networks must be made compatible with existing communications networks from an accounting traceability point of view. As a result, there are several accounting management issues that must be addressed in order for multimedia to be completely integrated into the standard telephone network. Finally, the network management features, and the essentials of operational multimedia systems, are of prime importance in terms of meeting user expectations.

Thus, the whole area of multimedia network management is in a state of transition, and its management is totally different from the usually accepted view of data network management. The various differences between the typical multimedia network and a typical data network are delineated below.

Quality

To begin with, as mentioned above, multimedia network management is concerned primarily with the applications and presentation levels of the OSI model. As will be seen within this treatment of the subject, the reception of a multimedia communication may indicate results that could be traced to either the transmission or the application program that created the material. If the quality of this information is less than expected by the receiver, more than likely, the root cause is with the applications program that generated the information to begin with. The emphasis, therefore, for the management of the system would be on the applications level of the OSI model. Any alteration in the processing of the multimedia information set should be reported to the network management system. This is so because the quality of the session conversations can be drastically altered by applications programs, and the network management function for a multimedia environment must track such alterations.

Media Integration

Second, the repositories for multimedia applications have a tendency to be fractured, with various storage protocols for different media types that have to be integrated for meaningful usage. The audio, video, and text all have different formats, coding schemes, and compression requirements, and therefore tend to be segregated from each other while being stored. Thus, each modality is inherently isolated.

Management Integration

Third, the term *multimedia* is merely a stepping stone for the ultimate desires of people to communicate with each other using all electronic means, not just voice. Each medium has its own unique set of measures as to how well its features and meanings are being sent and received during a conversational session, thus, the management of such intricacies presents special problems for the multimedia network system.

Technology Interaction

Forth, the type and variety of technologies associated with multimedia network management issues makes the total problem more complex than would be the case with data network management. For example, many technologies, both hardware and software, may come into play in order to create a multimedia presentation, and as a result the complexity of the network, and therefore its management, are more involved than might be the case for a data network.

Information Delays

The key issue for multimedia presentations is that the information needed for viewing must be made available to the presentation interface in sufficient time to reconstruct the image. The operative problem here is one of delay, where the segments of information needed for the presentation must arrive in time to be put to use.

Infrastructure Upgrades

Finally, certain multimedia support technologies require a considerable amount of network infrastructure in order to implement these features. Such multimedia support technologies as Asymmetrical Digital Subscriber Line (ADSL) require transmitting and receiving-end hardware to realize the benefits of these tools. The resulting network management problem is more complex than a typical data network configuration would be.

This book is intended for the technology professional who must understand the technical issues, architectures, and measurement parameters of the overall multimedia network and integration problem. The content for the chapters is summarized below.

Chapter 1 - Multimedia Systems Introduction

This chapter is an overview of the definitions, technologies, and systems that provide the multimedia experience. The various technical media components are defined, and their capabilities are traced through the various levels of integration to define conceptual networks and their requirements.

Chapter 2 - Multimedia Technologies: Audio, Text, and Images

This chapter discusses data storage technologies, on-line databases, data coding techniques, various software manipulation techniques, and compression capabilities used for multimedia implementations. Examples cited in the chapter include digital FAX systems, image processing, and animation facilities.

Chapter 3 - Multimedia Technologies: Video Systems

This chapter is a continuation of the previous chapter discussion with additional emphasis upon video capabilities. Various video processing techniques are discussed at length, including ADSL, HDSL, MPEG, JPEG, leading-edge analog systems, etc. The implementation hardware for these capabilities are then described, such as optical discs. Various types of compression schemes are compared for later use in establishing arguments for and against specific network management approaches.

Chapter 4 - Multimedia Technologies: Multimedia Data Management

This chapter covers the various issues related to the capture, storage, and usage of media presentation material. The current technologies and operational uses are discussed as they pertain to the special problems of multimedia data.

Chapter 5 - Multimedia Technologies - Transmission Systems

Various technologies have been developed to take advantage of the current trends towards fiber optics, such as frame relay and ATM, that have the capacity and switching capabilities to provide multiusers with multimedia bandwidth facilities. Other technologies have been developed to take advantage of current physical facilities and to stretch their bandwidth so as to allow them to carry narrowband video and imagery. These 2 types of approaches are discussed in this chapter.

Chapter 6 - Multimedia Technologies - Interfacing with Network Systems

Current network systems and their associated management facilities, such as the AT&T UNMA, IBM Netview, and others, are discussed in terms of their capabilities to support multimedia applications. The currently most popular LAN, Novell, is also addressed as to its facilities for handling such issues as delays, quality, and the other key issues associated with multimedia.

Chapter 7 - Multimedia Technologies: Coding and Compression

The various standards, such as JPEG, interframe coding, and MPEG, are discussed along with numerous compression techniques. Digital facsimile and audio compression techniques and standards are also discussed.

Chapter 8 - Applications Requirements for Multimedia Networks

The different operational elements of the typical multimedia network are discussed. The implications of the network management scenario are derived from the material covered in this chapter. The information loading and other pertinent parameters are then determined as a result of the preceding data. The most critical aspect of multimedia network management systems is how they interact with the established operational support systems within the telephone network. The generic operational support systems needed are discussed.

Chapter 9 - Multimedia Network Control Techniques

The other side of network management is discussed in this chapter, namely, the control issues. Additionally, quality measures are derived and described. The reporting function of the system is also described in terms of error requirements. Prediction techniques are evaluated, such as simulation, and Monte Carlo techniques. This chapter also delves into the various approaches to smart systems that could be used to control and manage multimedia networks.

Chapter 10 - Multimedia Network Management Architectures

The management approaches derived in the previous chapters are resolved into architectural concepts that can and have been turned into concept demonstrations in laboratory environments. How all of these requirements are integrated is discussed in terms of the OSI network management model. The various network configurations that support multimedia networks are discussed, as well as the connectivity issues associated with network management.

Chapter 11 - Multimedia Network Architectures

The various network transmission schemes that may be needed to implement the architectures presented in the book are explored. A breakdown of the various means for implementing network management of multimedia networks is presented and discussed. Finally, the differing protocols that can be called upon to implement multimedia network management are defined and discussed.

Chapter 12 - Multimedia Network Scenarios

Several multimedia applications scenarios are discussed at length in this chapter, from advanced automated teller machine banking booths to personal data retrieval kiosks. The network management implications of each concept are explored and architectures derived.

Larry L. Ball

Multimedia Network Integration and Management

Chapter

1

Multimedia Systems Introduction

Chapter Highlights:

The realization of multimedia capabilities within an organization depends upon the underlying techniques, technologies, and systems that comprise this unique extension of human communications capabilities. This book will address these various levels of multimedia integration, especially as they relate to the particular management of the networks that support such multimedia. Multimedia communication is the interchange of voice, print, and video among interacting individuals. Ideally, people want to communicate as if they were in close proximity to each other, even though they may be thousands of miles apart. And they want to be able to use all of their sensory and motor capabilities as possible. Herein lies the principal aim of multimedia communications. The interchange of information, mentioned above, and the forms that it takes is introduced in this first chapter. The actual interchange involves the speed, resolution, and integration of all modalities represented. The forms taken to assure this interchange include the physical media, the uses of compression and coding, delayed responses through storage and routing, and assisted use of the information available. Often overlooked, however, is the need for the management of the resources that provide these capabilities. This chapter will set the stage for discussions that explore these topics in further detail, in terms of both the enabling technologies as well as the management of these technologies.

1.1   Definitions: Multimedia, Hypertext, Hypermedia

1.1.1   Multimedia

The history of multimedia extends only into the recent past. It is essentially a current integration of the various techniques and technologies that can combine the existing media elements of voice, print, and video into a useful and flexible presentation format. More importantly, these various mixtures of media ingredients involved in communications delivery can be delivered by a variety of networking solutions that have been developed within the last 10 years. The constituents of the multimedia environment are reviewed in Technical Note 1-1. Several other technical advances preceded the widespread interest in multimedia, among which have been video teleconferencing. However, these applications have contributed to the development of multimedia, because they have stimulated the creation of new networking and applications capabilities, and such events are reviewed briefly here in this chapter, starting with the concept of multimedia itself.

One of the ways of describing multimedia is to define what it is not. The label *multimedia* has been increasingly applied to numerous products and services over the past several years, some of which have little to do with multimedia. For example, there are the various types of conferencing software products on the market; interactive data services, such as Compuserve, Prodigy, and America On-Line, slow-scan video teleconferencing; teletext services; high-capacity cable TV offerings; and even high-definition TV, otherwise known as HDTV.

The term *multimedia* has been overused and misused considerably, because it has become a buzz word and surrogate for many of the attributes of the *information superhighway or high-capacity entertainment services*. The *information superhighway*, for example, has been an ongoing initiative, for several years, aimed at putting the advantages of an increasing number of information services and databases into the hands of the average consumer around the globe. *High-capacity entertainment services*, on the other hand, address the issues of HDTV, increased bandwidth for coax transmission systems, wireless TV systems, direct broadcast satellites, video compression, and other such added capabilities for the conveyance of video.

Because the term *multimedia* has been used in various ill-advised ways, its definition has become corrupted and twisted so as to be recast as a cure-all for any telecommunications requirement or marketing scheme. For instance, recent TV advertisements have portrayed cable companies as adding fiber to their networks, and therefore providing multimedia capabilities for their customers because of the increased numbers of channel offerings to be provided. This, of course, is an appeal to choice, not to human-interactive, nor multisensory experiences that are provided with multimedia presentations.

Multimedia, of course, has nothing to do with the numbers of services provided, but rather with the way that information is provided.

Multimedia is an expression of communication realized as an interactive exchange of information where this exchange is conveyed in any of several different forms or in a combination of media formats simultaneously.

Technical Note 1-1
Basic Multimedia Components

Text information can be captured in 8-bit bytes, each byte usually representing a printable character, or the byte could represent some type of special or formatting character that is embedded within the text information stream. Each special character may represent a font type, font size, etc., which, in some cases, might be compressed for later retrieval and display. The issues of text handling are incorporated into certain well-established standards, such as ASCII, etc., and such well-known word processors as WordPerfect, Word, etc. More recently, Apple, Microsoft, and certain other companies have endorsed a new character representation standard that is 16 bits or 2 bytes long, which also accommodates other languages and alphabets, such as the Cyrillic alphabet.

Video software, which supports the collection and playback of moving images, performs according to a number of variable conditions. The video software providers allow the user to specify the number of frames per second that are to be captured, the frame sizes, compression algorithms to be used in the storage mode, and resolution of the images. Video is such a powerful medium that it is used in a multiplicity of ways in conjunction with the other modalities. For example, video can capture images, and when the frames are slowed or stopped, the images can be provided for analysis. Finally, slow scan video can slow down motion so as to provide the viewer more opportunity to study motion as it is progressing.

Audio information collection and usage requires digital conversion, compression, and collection controls. Since it cannot be seen, audio information has to have its own special hardware features included in order to be useful. The audio software also provides a selection of options, such as the sampling rate, sampling depth, and whether the capture is mono or stereo tracks. The new architectural concepts devoted to audio networking are making this medium flexible and more easily integrated into multimedia productions.

Therefore, even though it has nothing to do with the number of services provided, multimedia does, however, have something to do with the interchange of voice, print, and video information between humans, or between humans and various source repositories in an interactive mode of operation. It appears from this, and other examples that could be cited, that there are as many definitions of multimedia as there are appearances of the term in print. Next, we will delve into what multimedia involves.

The discussion of topics relating to multimedia must start with some appreciation for associated issues, such as the applications involved and the technologies required for multimedia applications. For instance, the term *media*, as used here, has a different meaning than its use in the entertainment realms of, say, TV, radio, or stage endeavors. In those arenas, the term *media* implies that each of those information or entertainment businesses operate as a separate distribution channel to the end users, and are, therefore, separate entities. Media experiences, as they are discussed here, are related to the senses, such as sight and hearing.

Figure 1-1 illustrates the various relationships between different kinds of media and their categorizations. If we are discussing a compilation of print (text) and graphics or imagery information, both of which types rely upon the sense of sight, we are dealing with *mixed media*. If we add voice and/or video information, we are now involved in what is referred to as a *multimedia experience*. Thus, multimedia is defined as the integration of voice, print, and video. Sometimes, the *multimedia* environment is a 2-way experience between 2 or more persons, or it can be an electronic conversation between one or more humans and a supporting database repository that can exercise several different communications modalities. Othertimes, *multimedia* is an interaction between a human and a database repository that contains the requisite imagery, audio, and text information needed to make the experience as rich as possible.

There are 2 other terms of interest in this area that have come into being, which are *hypertext* and *hypermedia* [9:41]. *Hypertext* is implemented in an environment consisting of multiple databases each with differing subject content, all of which are interconnected, and all serviced by some form of query software resident in a server or workstation. The hypertext commands involve requests to the software to assemble documents relating to the requested subjects from any of the connected databases. The command parameters include the subject, key words for the search, report length, possible organizational elements, constraints associated with the level of detail, etc. *Hypermedia* is similar to hypertext with respect to the overall architecture in that it is an extension of this concept, except that the content of the hypermedia experience also includes audio and video as well as text and graphics or imagery. These 2 concepts will be further amplified below.

The mistake of developing and using special-purposes machines in the mid-1970s to implement artificial intelligence (AI) applications taught the computer community about the error of trying to force special equipment onto each new technology that came along. Special applications were found to be more useful and adaptable if they were run on general-purpose machines accompanied, if necessary, by special-purpose software and hardware. So *multimedia*, as a beneficiary of this lesson, has been subsequently targeted for the general-purpose computer. As a result of new application technology needs, many of the previous features of the general-purpose computer were enhanced and/or expanded to accommodate these new requirements, while new capabilities were added to the general-purpose computer suite of capabilities as a result of general technological progress.

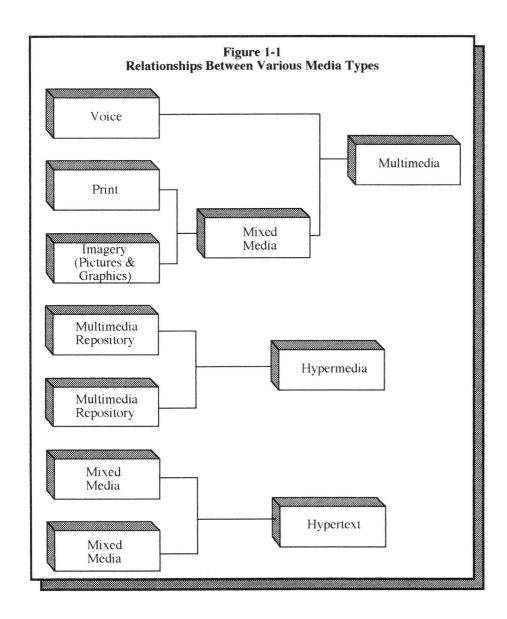

Figure 1-1
Relationships Between Various Media Types

Likewise, in a manner recognizing the AI machine debacle, *multimedia* is usually expressed physically as a set of technology devices or an integration of such devices that are fully capable of handling all modalities involved in the *multimedia* experience as well as other functions in lieu of having to use special-purpose devices in order to experience *multimedia*. The most common vehicle for multimedia, currently, is the suitably configured personal computer (PC), a general-purpose computer environment. In the context of the multimedia platform, there are several necessary features of the *multimedia* realm that have over time been more or less defined by the Microsoft Company and Apple Computer, as well as a cast of supporting companies.

The typical *multimedia* PC or Apple computer, acting either as a source for multimedia or as a destination, must be equipped, as a minimum, as follows, in order to support *multimedia* features [10:44-50]:

(1)   **CD-ROM and Controller -**
        A CD-ROM and controller is used as a source for the retrieval and display of high quality graphics, video, and/or pictures. The Microsoft-pioneered standard for the multimedia PC, known as MPC, includes a CD-ROM specification for the use of any special features associated with the CD capability. The CD-ROM will be covered further in succeeding chapters.

(2)   **Audio Input/Output and Digitizer -**
        A mono-track, audio digitizer and analog conversion plug-in board provides a source for audio data collection. It is capable of capturing sound at rates of at least 22 kHz in 8-bit resolution, and playback of these conversions at the same rates. Multimedia audio subsystems also have pass-through audio channels incorporated into their architectures. Digital signal processors (DSPs) can be used to filter, segment, and synthesize collected audio data. Audio collection is accomplished via pickup microphones, and is output to speakers that can interface with the system. These and other issues related to audio will be discussed in later chapters.

(3)   **Video Digitizer -**
        A digital video capture board is necessary for those multimedia applications where the operator is producing original multimedia presentations. A video accelerator board may be desired for use in conjunction with the playback of multimedia productions in order to smooth out and faithfully retain the normal speed of presentations. Additionally, a signal processor (DSP) is many times used to filter and sharpen video already captured. It is useful for both video and still image signals. With such a hardware capability, it is necessary to have a very sophisticated software package to take advantage of the hardware features to capture, or grab, individual frames, either on the fly, or at a rate reduced from the video delivery. Sound digitization is also handled through such a chip. This subject will be discussed in the next few chapters.

(4)   **High-Resolution Display -**
        A high-resolution (density), large pixel size (format), color spatial display area (information bearing) that is capable of rapid image

updates is required. Displays have gone through a series of resolution changes over the past 10 years. First there was the computer graphics adapter (CGA) and next the extended graphics adapter (EGA); finally, the video graphics adapter (VGA), standard has evolved. Within the current VGA standard, the spacing between adjacent pixels may be provided according to the manufacturer's technology implementation. The closer the spacing, the better will be the resolution. About the current best image that can be displayed on a VGA format is with the pixel spacing of 0.19 mm between the pixel elements on the display area. Tighter spacing between pixels provides for a better presentation of images and, it decreases the effects of aliasing, which is the jagged appearance of lines.

(5)   Random Access Memory (RAM) -
Computer systems now regularly have a special area of volatile memory set aside where applications programs and temporary storage are provided in order to enhance the operation of time-critical programs. This set-aside area is called RAM and is additional memory used for the purposes indicated above. It can range from as little as 2 MB to 64 MB. The use of multitasking, foreground, background processing requires that several programs be available at the same time for efficient use of the system resources. For example, video screening can be simultaneously operating in conjunction with background printing, while a foreground program, such as *Microsoft Word*, is being used.

(6)   Virtual Memory -
In addition to RAM, a system may designate a certain amount of the hard disk area, that is currently unused, as a virtual memory area for the temporary storage of programs and data files for use during real-time operations. The use of disk space is somewhat inefficient when video and audio playbacks are part of the operation of the application, but in some circumstances, its use is unavoidable if the amount of RAM is limited for whatever reason.

(7)   High-Capacity Disk Storage -
Multimedia workstations use a considerable amount of storage for database requirements. The typical multimedia system should be equipped with, at least, a 100-MB disk storage system. It is not uncommon for home systems to have half a gigabyte of storage capabilities. There are also external storage disks that can be attached to the main workstation that interface with the primary system via what is called an SCSI communications channel. Using

such an architecture, several different peripheral devices can be
linked to the main multimedia station.

(8)   Pointing Devices -
There are several approaches to the specification of the desired
place on the video screen for attention. Historically, they have
evolved from the arrow keys, to the mouse, to the track ball, touch
screen, and finally, may end up being embodied in brain wave
designation of the desired spot. Most common today are the
mouse and the track ball.

There are also several *ad hoc* standards associated with multimedia
activities that are also of value to the definition and completion of the
multimedia platform. These will be discussed in more detail later, but they can
be currently identified as [10:129-140]:

(1)   *MIDI* - a standard used for creating, networking, and playing back
audio and music devices;

(2)   *VISCA* - a chained and computer-controlled video system;

(3)   *Quicktime* - a movie-capture software program used on the Mac
systems; and

(4)   *Video for Windows* - a movie-capture program, similar to
Quicktime, for the PC.

Although *Video for Windows* is the *de facto* video standard for PC platforms,
Quicktime has received more support from the user population, and versions of
Quicktime have been developed and delivered for the PC market as well.
      Quicktime is a standard for the delivery of multimedia programs. In the
Macintosh world, it is called an extension, which means that it is a utility
program that is resident in a special location in the system software area.
Quicktime can be used to store movies in, or associate movies with, other well-
known Apple-related applications, such as *Microsoft Word*, *Microsoft Power
Point*, and *Claris Hypercard*, to name a few. It is a digital capture, storage, and
display program that has its own formatted output that, in turn, gives it its own
standard appearance.   *Quicktime* is also being modified to support video
teleconferencing facilities.   This development capability will likely make
Quicktime even more widely accepted as a video standard.
      Although not standards *per se*, certain other word processing
applications programs are viewed as almost indispensable for generating
multimedia applications, such as *Microsoft Word*, *WordPerfect*, etc., that have
built-in capabilities to include video clips, graphics and imagery, audio, and, of

course, text information. For applications that require animation, there are numerous sophisticated programs currently available that can satisfy such needs. Additionally, various compression algorithms, some of which are also *ad hoc* or established standards, are discussed under such titles as ADSL, JPEG, and various Apple algorithms or Microsoft programs. All of these tools contribute to the realization of multimedia products, but must be integrated appropriately to achieve the desired results. As a result, the management of a multimedia network is very dependent upon how these tools are treated and handled in each application.

When 2 or more media are integrated for usage across 2 or more separated storage locations, the opportunity arises for a broader range and richer depth of information integration covering some particular subject. These types of situations give rise to such conditions as networked multimedia environments and different types of applications. Networked multimedia environments will be discussed in more detail later. The richer blend of possible applications for multiple media storage involves certain technologies and implementation issues that are referred to as hypertext and hypermedia. These will be discussed below.

1.1.2   Hypertext and Hypermedia

Hypertext is an evolutionary extension of multimedia technology. The overall concept is illustrated in Figure 1-2. It involves a step beyond the hierarchical storage and retrieval systems of the past, where direct references to subject matter could be extracted and compiled into a document form for the requester. The approach is analogous to relational database searches, where the subject material is pulled directly from stored records and then assembled into readable form for the user. This latter feature involves intelligent capabilities within the software. Expert systems are used in conjunction with hypertext and hypermedia applications to aid in the assembly of such documents as requested by the system user.

The whole process involved in the utilization of hypertext technology can best be appreciated by thinking of an authoring system that operates as follows. The activity consists of the creator of the document clipping paragraphs from several different articles, rearranging them, and then pasting the reorganized grouping onto several pieces of paper in order to assemble the final document. The articles are analogous to different information files located at different locations as represented by the different magazines and/or papers from which the articles are derived. The most important technology aspect of hypertext is the integration of these disparate sources into a coherent and cohesive rendering of text, graphics, and imagery into an essay devoted to some specific subject. Technical Note 1-2 reviews the essential features of hypertext.

Different hardware and software database systems may provide varying types of text, graphic, and imagery to the applications platform via the different formats employed. At this point, the integration criteria are applied to the selection, segmentation, cropping, and clipping of the various media materials for reassembly into the required and completed document. This completed product may also be stored in electronic form.

Hypermedia, on the other hand, is an integration of hypertext with additional media inputs from audio and imagery or full-motion video. Hypermedia is yet another level of an applied experience for the user, where various combinations of media clips containing voice, video, still imagery, and text are all integrated into a multimedia presentation scenario. The same scenario applies as for hypertext, in that various sources of information, in different media formats, are assembled by the computer software to yield an essay or dissertation on some selected topic using all media available for incorporation into the completed package.

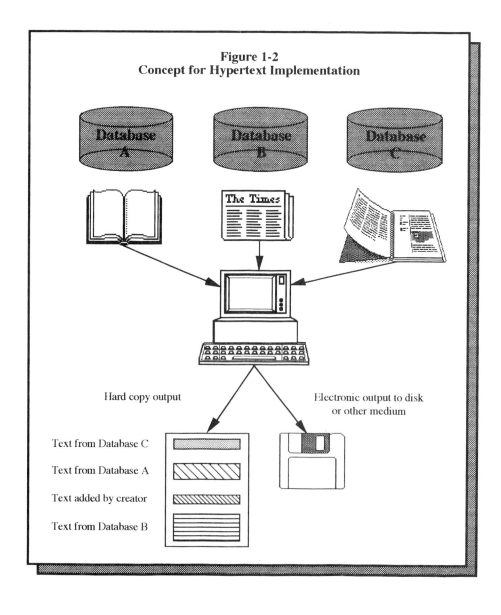

Figure 1-2
Concept for Hypertext Implementation

The various pieces of information used are retrieved from a specially designed database, or from a group of different databases, at which point integration may be performed by an expert system software program [7:21]. Both of these technologies are explored at length later in this chapter. Figure 1-3 illustrates the hypermedia application. Various databases are accessed, just as with the situation surrounding hypertext, except the modalities are different media. These media are researched and parsed as needed to satisfy the requirements of the specific request made by the user. The request is also multifaceted in that it may indicate the subject, length, and other compilation parameters. The output may be in any of several forms depending upon the needs of the requester.

Technical Note 1-2
Hypertext Composition

The key to hypertext is its software. The composition of a hypertext document requires several features, among which are:

(1) an interface for the user to specify the topic desired, length of paper, source information, possible generic outline, etc.;

(2) software that will search databases, extract the information required, and reassemble the text from the specifications given; and

(3) access capability to databases that can be used to provide the source information needed.

In the meantime, it may be concluded that hypermedia is implemented in a distributed media database system, where multiple data repositories store different modalities and can be accessed from several different locales [8:54]. Each workstation involved has software capabilities required to take requests for hypermedia tasking interfaces and to produce these products as one or more multimedia scenarios that can then be packaged for viewers.

The processing system software takes the task assignment, which can be expressed as simply an assignment to produce a multimedia scenario, which may be defined as a subject, a length, facilities available on the workstation, and any search constraints set by the operator. The applications software will then, based upon its network relationship with other information repositories on the system, query all such sources available, extract information that may fit the constraints of the project, and generate the production by linking multimedia clips together in some logically ordered sequence. This capability, in its final form, requires an intelligent backend to assemble the collected raw source information.

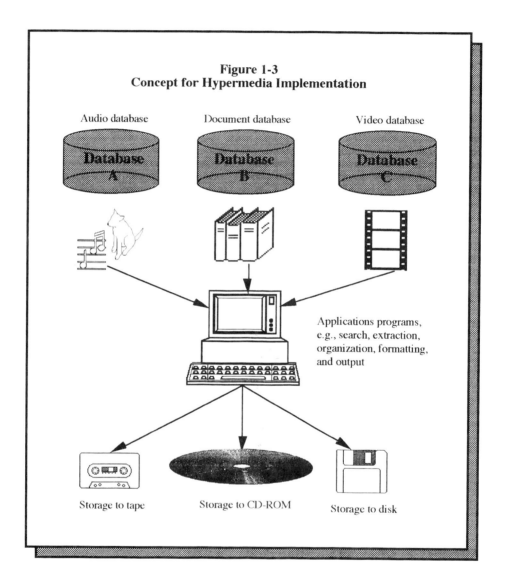

Figure 1-3
Concept for Hypermedia Implementation

1.1.3   Personal Multimedia Systems

There are a group of other technologies that have given rise to an area of applications development known as personal multimedia systems, some of which are described here. The information age focuses upon the creation, access, and usage of information for the benefit of the individual, in fact it is all about personal information access. One need only consider the Internet to appreciate this fact. Distributed processing, inexpensive personal computers, and high speed transmission systems all support the notion of ease-of-access for the individual. The personal aspect of personal multimedia systems implies that the delivery of services for the individual must be accessible by all, and tailored for each individual's needs.

Therefore, the thrust of multimedia is the promotion of personal interaction, whether it be a conversation with other humans or information gathering from an information server. The preceding discussions of hypertext and hypermedia are good examples of how the personal issue has become a central theme for various types of media conversations. This aspect of personal involvement is being explored both in the development laboratory and in commercial offerings, in such concepts as personal communications systems, personal directories (to be discussed in detail in Chapter 12), personal manufacturing, and personal newspapers and magazines, to name a few.

For example, at least one company already offers personal newspapers using such sources as *The Wall Street Journal* to compile the information of interest to the individual. This electronic version of the daily printer's ink and recycled paper offering uses a personal description (personality profile) of the reader to segment, i.e., sort, partition, and augment the stories of the *Journal* that would be of interest to the reader and to provide additional and more in-depth coverage of the subjects desired. The software examines the user's profile and matches that with news articles matching the profile. The interpretation of what constitutes a match between profile and news articles is the object of the expert system utilized. The initial subscription rate for this service is to be $12.95 per month. It appears that part of the cost will be paid for by advertising revenues where advertisers want their ads put in front of the readers intermingled with the personal newspaper articles.

The personal directory, on the other hand is a concept that calls for the periodic consolidation of the subscriber's calling and called phone numbers for use by that subscriber, either via print-to-voice or other means. The concept was explored at NYNEX in 1989 by Nathan Felde and this author. It calls for the collection of user calls and calling numbers using the automated message accounting (AMA) systems used in central offices for the detailed call accounting of all incoming and outgoing calls. These numbers are then reversed to obtain the names and addresses of the calling and called numbers and the directory is then assembled from these inputs. Print-to-voice or print-to-tone technology can then be used to dial numbers or inform the subscriber of the number to be dialed. As with the personal newspaper discussed above, a percentage of the costs of the directory is supported by advertising revenues from vendors that want their ads put in front of the subscribers to the directory.

Internet and the other on-line computer data services are about to introduce browsing capabilities for searching and delivering user-specific information. These capabilities are to be intelligent in the sense that they will provide expert systems backbones to achieve the basic needs of finding, compiling and reporting facilities. These systems will then begin to deliver the hypertext and hypermedia services that were previously discussed. Hypermedia services have already been demonstrated in the laboratory at such places as the NYNEX and Bell Atlantic Media Labs, and are now in the process of being rolled out into the commercial markets by commercial vendors. The media capabilities being rushed to the market by Compuserve, Prodigy, America On-Line, etc., parallel the services described above and are the beginnings of the hyperspace era.   Many developers are also now providing video capabilities to the Internet so that the user will have access to multimedia presentations.

The concepts identified above are rapidly becoming a reality.   These developments and their commercial versions will probably, at some point, be integrated. This integration of the personal aspects of multimedia will also be reviewed in Chapter 12.

1.1.4   Networks and Integration

The issues associated with multimedia networks and integration can be manifested in many ways [2:20-40].   Their complexity ranges from a single workstation as a self-contained network, to a complex network of many workstations operating in conjunction with a variety of other special devices, such as scanners, video cameras, video juke boxes, etc., all joined through the facilities of local networks interconnected with wide area services consisting of either broadband transmission facilities or, compression technologies that increase the utility of narrow band links. As will be pointed out periodically throughout this book, the management of multimedia networks will be supported primarily by the definition, capture, and evaluation of quality status information about the multimedia information passing within the confines of the network under discussion rather than the evaluation of outage and congestion information related to the transmission and networking layers of that system.

From the standpoint of the multimedia network as a collection of different devices serving some useful purpose, the multimedia network is not much different from the usual data network. However, the multimedia network is not driven by the exactness of the data formatting nor its contents, but rather by the degree to which the user can interpret information passed over this assortment of hardware, transmission facilities, and software functionality. In short, the emphasis on the management of the network functions has, with the multimedia network, passed from physical and low-level functional monitoring to the higher levels of applications monitoring and processing, e.g., Layer 7 of the Open Systems Interconnect (OSI) model.

1.2   Overview of Essential Technologies

1.2.1   Enabling Technologies

As with any other application, multimedia is implemented by certain enabling technologies suitably integrated to achieve the required results [1:5-20; 3:30-42; 4:109-113; 5:1-6]. One of these is the communications network, another is the integrating network software needed to create the various multimedia productions, yet another is the hardware needed to facilitate the features required, and finally, there are the storage media that contain the information of various types needed to convey the multimedia message. Each of these will be introduced below and expanded upon throughout this book.

The communications network component of the multimedia problem represents a networking challenge. The various media that are being transferred back and forth across this system are conveyed at different rates, and from many sources to one or more integration stations, all almost simultaneously and interleaved. Some of the information is real-time, as with teleconferencing, while other information transfers are concerned with non-real-time information, as with stored animation or video clips. There are various approaches to communications interconnection, but the one that makes the most sense for simultaneous multimedia applications is that referred to as the *packet switched networking scheme*.

Technical Note 1-3
Packet Switching Technologies

Packet switching is a technology that has been around for several years and consists of segmenting information into a series of chunks, packaging these chunks into formatted frames, complete with addressing and control information, and routing each frame or packet to its destination based upon the transport protocol defined for that system. Some of the common data protocols are examined later in this book, but it is worth mentioning at this point that any given multimedia network will have to operate such that there are multiple protocols in use simultaneously across the net.

Routers are devices that provide connectivity and compatibility between local area networks (LANs) or other network architectures. This compatibility feature means that networks, working with dissimilar transmission and packet protocols, can interconnect if some type of software conversion between protocols is provided. Unfortunately, most routers can only support transfer rates of about 2000 packets per second (pps). The rate required for any type of video transfer is at least 2400 pps.

Packet switching is elaborated upon in Technical Note 1-3. Video and audio packets are then sent out through the broadband pipe, which in future cases will probably be microwave or fiber optics links, either in a time-slotted fashion, code division, or frequency-multiplexed manner. These packets are received at the termination point and decoded by analyzing the overhead fields for destination, error coding, and other operative features that assure the proper delivery of the information contained within the packet.

The integrating software for multimedia functions may be based upon intelligent software concepts. Such techniques are based upon organizational features of the software, such as summarizing, cropping, and scaling. Other techniques, such as inferential methods, allow the system to identify specific objects or topics of interest.   The media control is managed via a media operating system that is summarized in Technical Note 1-4.

Technical Note 1-4
Media Operating Systems

The multimedia integration software is essentially a network operating system. It provides the environment for both internal and external interfacing to multimedia repositories. The principal method for the external communications with outside media databases is via packet technology. The media network operating system (MeNOS) performs the following essential tasks:

(1) arbitrates between internal and external requests;

(2) recognizes one or, preferably, more communications interchange protocols; and

(3) provides the capability, or hooks, for the insertion of network monitoring and management features to be located within the media network environment.

Specifically, the media software operating system has to be designed and constructed so that the following features are provided:

(1) Video and data packets must be provided to their destinations on regular time intervals so that audio or video pixel dropouts will not be incurred. This concept is referred to as *isochronous data flow*. *Isochronous data flow* may only be realized by employing fixed length packets in a highly congested network system.

(2) The network operating system controlling the media packet flow must be capable of managing the data flow such that the information packets create as little overhead for the system as possible. This may entail minimizing error detection and correction checks so as to minimize packet retransmissions.

(3) The packet system must also minimize overhead by making the packet management system as efficient as possible in order to guarantee a high throughput of packets. In today's sophisticated networks, a certain minimum overhead per packet is required, such as the packet header, source address of the packet, the destination address of the user (referred to as end-to-end, or transport level addressing), and error code.

The operational mission of the system is to link video and still image clips while also associating audio with these clips for a total cohesive presentation where a multimedia effect is required. The software has to provide a wide range of processing features in order to support the options needed for such a complex set of requirements. Depending upon the specific software capabilities, the system performance may react sub-optimally to perceptions of acceptable audio and video quality. Certain software programs used for multimedia productions may not operate at peak functional capacity unless the hardware is appropriately configured. Thus, the best performance results where there is an effective operational combination of both hardware and software capabilities.

The architectures that support the transmission of video clips, audio tracks, or text documents segment these various media divisions into packets for transmission. Each media packet is then packaged into a form suitable for conveyance to its destination. Each packet is assigned a header for routing purposes. This architecture is illustrated in Figure 1-4. The key is whether the aggregate number of packets for each media division can be reassembled and reconstituted for display in time sequence given the delays inherent in a packet transmission system. This question will be addressed later.

The hardware systems needed for multimedia technologies have to be comprehensive as well. There are several features that have been developed to boost perceptible performance. The major hardware features of an effective multimedia system are outlined below.

(1) **RAM -**
Random access memory (RAM) is resident memory allocated for the execution of the operating system, programs, data files, buffers, etc. In a multiprogramming environment, several programs may be resident in the main memory RAM at one time. The operator may toggle among each of several programs resident in RAM, as required, without having to go through the old steps of calling a program, using it, and then closing it, when another program is needed for execution. System RAM options now provide up to 64K of memory or more for such purposes as those defined above.

(2) **Cache Memory -**
Another accommodation provided by operating system developers some years ago was the step to allocate part of the available RAM as a temporary repository for digitized audio or video captured from outside sources, or as a temporary storage for animation scenarios. In order to implement this feature so as to maximize all of its power, hardware manufacturers started to increase the size of the RAM available, or optionally available, for the use of temporary storage. The idea behind cache memory is that data, collected from either outside sources or retrieved from peripheral disk sources, must be

either collected at real-time rates or read out from existing storage at real-time rates. As a result, and in order to take advantage of this increased capacity, software processing programs became more sophisticated, with more realistic capture and playback characteristics.

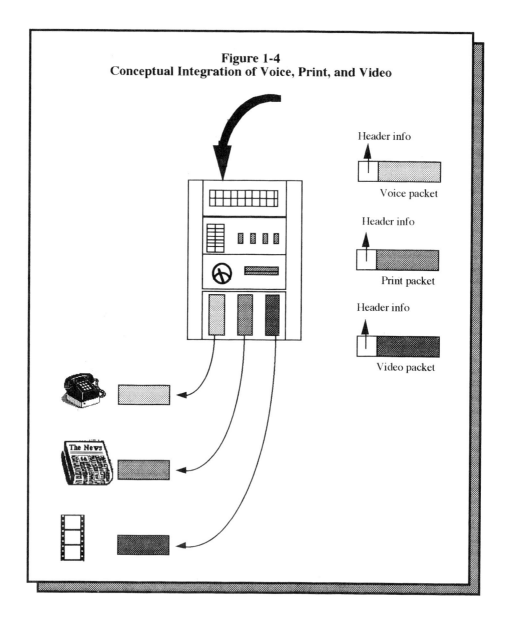

Figure 1-4
Conceptual Integration of Voice, Print, and Video

(3)   **Addressing Features -**
As the sizes of RAM, cache, and disk memory increased, the addressing requirements for these memory slots also increased.

Eight-bit addressing of the early computers gradually gave way to 16 -bit addressing, and within the last couple of years, systems have begun to utilize 32-bit addressing schemes, either optionally or as a standard feature of the systems in which this feature resides. Disk memory now runs into the gigabyte ranges, with terabyte and higher capacities soon to follow.

Table 1-1
Multimedia Technologies

Technology Area	Technology Description
Transmission Technologies	The transmission technologies range from narrow-band analog and digital signaling over twisted-pair wire to high-speed packet systems carrying multiple conversations over the same network simultaneously.
Storage Technologies	The storage technologies are concerned with media database management capabilities that include storage and retrieval protocols, data/image compression, and high-capacity hardware systems with multiple access and playback features.
Applications Technologies	The applications technologies include applications status monitoring and reporting algorithms, production systems, such as animation and/or media integration capabilities, and the like.
Collection Technologies	The collection technologies revolve around collection devices, such as document scanners and video input, integration of multiple informational modes of communication, human interface devices that send and receive such integrated information and parse it for the appropriate presentation medium, and interfacing capabilities.

1.2.2   Implementation Technologies

Since multimedia is implemented using any number of features and facilities, the technologies involved tend to be considerable in number. The key capabilities that are required for multimedia interactions can be segregated into transmission, storage, collection, and applications.   Table 1-1 shows the explanation of these technologies associated with multimedia.   In this introductory chapter these essential capabilities will be briefly explained, and in later chapters these features will be discussed in more elaborate detail.

Technical Note 1-5
Database Systems Terminology

Database systems can be classified as either centralized or distributed, and either hierarchical or relational. A centralized database is one from which all storage to and retrieval from resides. A distributed database is one where there may be two or more databases within the system, each having storage and retrieval access. Beyond this the distributed database has many additional features and restrictions in terms of how data is updated, how retrieval is exercised, duplications of database elements in two or more user locations simultaneously, and so forth. The exercise of access, update, and storage rules are specific to the particular product being used.

The concept of a hierarchical database system can best be appreciated by thinking of entry into or exit from the database as being through only one door or root. This presents some time-consuming situations if a search is conducted to retrieve information about attributes over many different records. In any such database system, a specified search method is employed that will traverse all records in the database according to some preset algorithm. For a database that has 1000 records, for example, all 1000 records have to be searched, using any search algorithm, in order to fetch the sought-after attribute information, providing the information is spread out over all 1000 records.

The concept of a relational database system can be thought of as having several different doors for entry. Thus, if the relational database is queried for some information about a specific attribute related to the system, such as the cumulative availability of some item over the previous year, this information becomes instantly available without having to examine each record containing such information. This instant availability occurs if that attribute was previously set up as a key field in the relational system for purposes of retrieval. Thus, all occurrences of the item of interest are stored in a special record or record set, and its availability can be accessed by examining one or a few records specializing in that data. These various technologies are discussed in more detail later in this chapter.

Video, imagery, and audio information is retained in hierarchical databases, because relational systems are inappropriate for such uses. Also, centralized storage repositories are more likely to serve the needs of the media requirements better than decentralized ones, because there is little use for dispersed information unless storage at one location is an issue.

1.2.3   Multimedia Database Management Technologies

There are a variety of storage techniques and formats used in the 3 multimedia modalities of voice, print, and video. These methods have been developed to satisfy their own needs, and some of these will be briefly discussed here. Database systems have evolved from early forms of data classification to those of today. Technical Note 1-5 provides a look backward into this initial form of data organization.

The Apple Macintosh environment has, for several years, had the facilities necessary to collect, store, retrieve, and paste imagery, both still and video, into documents. Additionally, the Apple environment, which amounts to a media operating system, has the ability to access and integrate various media into the same storage platform. The major drawback for the Apple media interchange is its slower information exchange rates, which are about 240 Kb/s.

Unfortunately, the PC environments have lagged in providing such capabilities. On the other hand, alliances among certain telecommunications and computer providers, such as Kodak, Novell, and Lotus, have fostered the developments of image services for such products as *Lotus Notes* and for LAN operating systems such as the Novell *Netware* system. The Novell product is called *image-enabled Netware*. This added feature is essentially an image handling operating system, that falls short of the video handling capability needed for multimedia situations. Table 1-2 shows the 3 major services provided by the *image-enabled Netware* capability. Some of the more popular data management systems in use today are discussed in Table 1-3.

Table 1-2
Image Handling in the PC Environment

HCSS   High-capacity storage services (HCSS) provide image accesses between the workstation and the high-volume storage units, typically optical jukeboxes. Optical jukeboxes are capable of holding at least 1 Gb of imagery and other media per platter, where the total system may contain up to 100 platters.

DMS   Document management services (DMS) provide a capability to organize documents into folders and workbaskets. Icons represent the various documents on hand, and can also be used to facilitate the integration of documents into the creation software.

IMS   Image management services (IMS) allow access to and control of the image files themselves. The IMS keeps track of the formats, and interprets images prior to their being sent to the requesting workstation.

1.2.4   Data/Image Compression Technologies

Compression is the process of reducing the amount of symbology, at the transmission point, necessary to replicate a given piece of information at the receiving end [1:15-19]. Compression is based upon the principal of entropy, where the theory is that there is a minimum number of bits that can replicate a given alphabet. Some compression techniques are lostless, meaning that they adhere to this principal of entropy, while other compression techniques allow some loss and therefore do not reproduce the original bit stream exactly. Compression systems are likely to be one of the cornerstones of multimedia for the foreseeable future for many reasons, one of which is that compression conserves storage space, and another of which is that transmission bandwidth can be better conserved through compression. Compression algorithms were made part of the FAX standards specifying Group 3 and Group 4 protocols, and compression algorithms have been key to data packet systems implementation for years. There are many different kinds of compression algorithms that will be discussed further in later chapters.

1.2.5   User Platform Technologies

The multimedia platform is an elaboration of the current computer workstation environment where all sensory input and output capabilities are controlled from one station. The beginnings of this new era of information flow are already in view where multimedia workstations are being advertised that embody voice input and output, as well as video display, and the usual display and keyboard features. Such systems incorporate FAX, telephone, TV, and various data services currently known under such names as ISDN, etc., in addition to the expected applications that exercise these capabilities. The MPC, *de facto* multimedia standard, has, if nothing else, set a baseline against which applications and utility programs can be developed.

Table 1-3
Typical Data Management Systems

(1) IMS - This is a hierarchical database retrieval software package that was used on large computer mainframes. Hierarchical database management systems are relatively fast for tree search types of applications, but slow for category types of searches. Images are stored in compressed or uncompressed files, and are inherently tree structured in nature. However, header information in each image file may contain reference information that could be used to perform category searches. Such information might include time and date stamps, and key word inclusions provided by the author.

(2) Oracle - This database capability is a relational database management system originally developed for large computer systems, and later adapted for networked workstations, and eventually PC networks.

(3) SQL/DS - SQL/DS is yet another relational database management system that has become a standard against which other data query systems have been measured. SQL, or something like it in functionality, could be valuable in both hierarchical and relational image searches.

1.2.6   Transmission Systems Technologies

Multimedia must take advantage of all sorts of transmission systems, both physical and functional [4:4-8]. Some of the more interesting or important ones are outlined briefly below.

(1) **ADSL/HDSL -**
Recent developments in coding and modulation techniques have given new life to twisted-pair copper. Some of these innovations are known as ADSL or HDSL. ADSL is a technology aimed at providing compressed video at 1.528 Mb/s in one direction plus a 16 Kb/s signaling channel in the opposite direction. It was developed to support the telephone companies in their efforts to create transport facilities for the delivery of movies to the home as part of their positioning to compete with the cable TV (CATV) companies. HDSL, on the other hand, was developed to provide a low-cost 2-way video teleconferencing capability.

(2) **ATM/SONET -**
Several packet switching technologies have emerged over the recent past that provide transmission and data control over the first 2 layers of the OSI model. One of these is known as frame relay, but it has been leapfrogged, in terms of potential utility, by a second, known as asynchronous transfer mode (ATM). Frame relay is based upon a variable length packet, while ATM is a fixed length packet system. For this reason, ATM is thought to be more suitable for multimedia transmissions. ATM is a packet switched networking system technology that takes information packets that have been parsed and packetized at their originating points, and then repackages them by the ATM protocol to include a header for each packet that designates the routing information needed for each packet. ATM and other high-speed packet switching technologies, which are implemented via fiber optic systems, require a suitable interface between the fiber and the electronic switching. This interface is known as SONET and comes in several speeds that will be explored later.

(3) **Coaxial Cable -**
Coax, as it is called, has the ability to carry a wide bandwidth of information. Coax has been used almost exclusively in community-access TV systems, known as CATV. The systems that have evolved around this technology, and its supporting electronics capabilities, are 1-way, rather than interactive, and have to be reamplified at relatively short intervals, because the attenuation

associated with coax is significant. Most of the repeaters used in CATV systems amplify in only one direction, and therefore, signaling back to the servers and headend equipment is not possible. Typically, all selections have to be presented to the termination points at the user end, at which point the user can make his/her own selection using some type of demodulator. Two-way repeaters have been installed in some CATV systems and upstream signaling implemented. However, these implementations may not be practical from a cost point of view.

1.3   Classifications and Standards

1.3.1   Hierarchy of Multimedia Creations

This chapter, in fact this whole book, is about the discussion of different types of information and its utilization. Anything that humans can simultaneously or interactively exchange, such as pictures, language, signs and symbols, etc., could be classified as multimedia when these media are used interchangeably and coincidentally with each other to create the communication desired. The information needed for multimedia exchanges comes in many forms, and can be defined variously. For instance, information such as that used to compile multimedia conversations or information retrievals can be hierarchically identified as follows.

(1)   **The Message Medium -**
       The message may be a series of interactions between sender and receiver, using signs and symbols, such that a point is made and/or explained. This message may be lengthy or a set of lengthy combinations of video, text, and audio clips that can be used to create the desired impression. The medium over which the message is conveyed is key to the impact of that message, because video interchanges require sufficient bandwidths to convey the content.

(2)   **The Message Format -**
       The organization of the multimedia document consists of the definition of thoughts and impressions, the grouping and sequencing of information, and the identification of points to be made. This formatting activity allows the sender to create and organize the media to be used in the message, by segmenting and sequencing the order and content of these facts and information items.

(3)   Message Elements -
Data elements are the lowest level of media interchanges. They involve the identification of needed facts and information, as well as the development of conclusions. The overall system organization contributes to the data organization by providing the database architecture and storage formats. Both the data elements and the system organization contribute to the definitions of the specific data elements.

While this lexicon for media does not actually define multimedia, it does help to organize our thoughts about one of these important modalities that are involved in multimedia [1:20].   Hypertext and hypermedia are 2 of the contributing modalities to multimedia. Table 1-4 describes the role of hypertext as a related issue for multimedia development.

Table 1-4
Documents and Hypertext

Documents are collections of text and visual images that are organized according to format restrictions to represent some thesis. *Data sets* are organizations of information types and their contents to provide the building blocks for subsequent documents or collections of system information. A conglomeration of signs and symbols that gives complete treatment to a particular subject is a system organization of that subject.

The use of the term *document* is meant to imply graphics or pictorial information as well as text only. Documents may further be classified as being linear or nonlinear. Linear documents are those that are intended to be read from first page to last in page-by-page sequence. Nonlinear documents are those that are intended to be read in skip fashion, i.e., the reader is required or encouraged to read pages and topics without following any obvious sequence. Hence, nonlinear documents are more like data files made available through a server structure, such as a database. A physical demonstration of a linear document might be a mystery novel, while a nonlinear document could be an encyclopedia.

Hypertext is the generalized term applied to nonlinear documents, and the integration of material that results in the hypertext output. *Nonlinear* here means that the hypertext document is assembled from an indeterminate number of source materials of variable numbers and dimensions. Print has, until the advent of the phonograph and the camera, still or motion picture, been the principal means of conveying information between conversationalists. It is a very powerful medium, principally because it is the basis upon which civilization has made itself known.

This medium is very powerful because of our ability to paint images with words and to instill ideas and concepts with words. A good example of how images can be envisioned using words, printed or spoken, is the use of radio to convey lifelike characteristics for otherwise inanimate objects. Radio and TV commercials abound with talking dogs, brooms, toilet bowl cleaners, and the like. More recently, textual documents have been enriched by the addition of other media into their contents. Many word processors now have capabilities to incorporate audio, imagery, and full-motion video clips into their contents.

1.3.2   Categorization of Media Effects

A similar analysis of the other media provides the categorization of their attributes from which combinations of multimedia capabilities may be constructed.

Imagery carries the impact of conveying impressions that otherwise would require many words to deliver and, sometimes, in a way that words have trouble describing, unless the writer and reader have exactly the same understanding as to what each word means. Imagery and graphics are different in that each contributes something different to the reader's impressions. Imagery conveys reality, but sometimes reality falls short of delivering the entire message or the exactness desired, while graphics can show accuracy which cannot be captured by the photographic device.

Vocalizations provide an additional element of personalization to the transfer of ideas and messages through the use of hearing. Spoken language adds the various aspects of personalization, e.g., emphasis, emotion, and intonations. In a multimedia environment, voice accompaniment may be part of the video action or used to describe it. It may perform text-to-speech conversions, or it may provide background descriptions of still imagery. Spoken language conveys the personal touch, and as such adds a totally new dimension to the exchange of language. Audio segments can be categorized by the data sampling protocols to which they are subjected.

Finally, the video component of multimedia adds the aspects of space and time to the conveyance of language. Even though video is a recent innovation, its impact upon the viewer has been enormous. It conveys, as no other modality can, the complete replication of people as they move about in their many activities. The video experience may be full-motion, flicker-free, or something less, such as slow-scan TV, where the motion appears to jerk as it progresses. Such impediments are prevalent in narrow band video conferencing systems. Video, as a component of the multimedia conversation, has been the source of much experimentation, and appears to be very effective, in terms of creating an impression, when inserted into the situational presentation. Thus, a presentation might contain windows or areas on a screen that display text, graphics or pictures, and video information, when transmitted.

1.3.3   Multimedia Data Standards

There is a significant degree of importance associated with the organization of multimedia data as it relates to the management of networks that are managing such issues. Each of the 5 pillars of network management relies upon multimedia data as follows:

(1)   The fault management algorithms will, at a minimum, need to be aware of the sizes, formats, and other internal features of imagery and related files in order to analyze the source and nature of possible problems.

(2)   The configuration management features of network management systems must be aware of the special requirements of multimedia transmission and playback of media presentations in order to assess whether fault inquiries might be within the domain of the equipment and/or software inventory, or whether the problem may be the result of system outages.

(3)   The performance management module of the network management system must be aware of the index features of the image and other related files of the media presentation in order for transfer and usage parameters to be assessed and evaluated.

(4)   The security network management module must have the ability to recognize the designated recipient and to verify the proper receiver. For video and imagery, the security must be applied so as to minimize overhead requirements for verification.

(5)   The accounting management module must be capable of tracking usage in terms of total volume, sessional usage, and by modality, e.g., audio, video, and text.

Advances in the arenas of standards and technologies as they pertain to multimedia databases have made the advancement of special media databases more practical. Explained below are a few of these more important and recent advances.

(1)   **CCITT H.320 -**
This is the codec (coder/decoder) standard for ISDN. It allows for variable presentation formats and specifies the coding mechanism

for the transmission of video teleconferencing information between points of usage.

(2) **CCITT H.261 -**
The H.261 standard describes the transmission of video in terms of the picture elements (pixels) and lines per frame. This standard is an attempt to create interoperability among different manufacturers, and to reduce bandwidth requirements for differing frame sizes and frame rates. The standard defines 2 different frame resolution formats, one for 352 pixels by 288 lines per frame, known as the common interface format (CIF), and the other being the quarter common interface format (QCIF) consisting of 176 pixels by 144 lines per frame. Both formats can support frame rates up to 30 frames per second.

(3) **MPEG -**
The MPEG standard was conceived and developed by the International Standards Organization (ISO) as a motion-compensated interframe coder-decoder (CODEC) system. It was originally targeted for videophones and video teleconferencing systems, but has been expanded to include some of the copper transmission technologies that are intended to compete with certain features of the CATV industry offerings. The bandwidth requirements for the MPEG I and II versions currently published range from 1.2 to 10 Mb/s. The uncompressed video for a National Television Standards Committee (NTSC) signal is 80.6 Mb/s; thus, the MPEG standard seeks to reduce the bandwidth by a factor of 67:1 for a 1.2Mb/s signal, and about 8:1 for a 10Mb/s signal.

(4) **Compression -**
There are a number of compression algorithms that have been created to allow minimal storage space for media products. These compression schemes, such as JPEG and Stuffit, will be discussed in more detail later.

(5) **Formats -**
Formatting standards for text have been in existence for years, while the formatting for audio and video are much newer needs. Video formats have been pioneered by Apple Computer with TIFF, and other formats have emerged, such as PICT and GIF, as well as about 25 others being used in the typical network management system. Their numbers dictate that the required format readers be available for use with the applications software in use.

(6) Header Data -
Indices associated with each media file provide valuable
information about the contents of the file, its creation and update
characteristics, and carries other information that provides answers
to issues of size and importance.

1.3.4 Organizational Forms of Multimedia Data

Data organization in multimedia is concerned with the associations of different
forms of information with one another, how this information is stored, how it is
retrieved, and its compatibility within various usage platforms. We will explore
each of these. Data storage systems for textual information have received much
more development attention than those for graphics, still frames, or video. A
brief look at the history of data repositories is provided in Technical Note 1-6.

Technical Note 1-6
Historical Background for Data Repositories

The data repository systems and structures for multimedia accesses were
originally developed for data storage and retrieval transactions, all of which
were text character-oriented. At these times, information that needed to be
stored was entirely text. Groupings of text are still arranged in files and can be
accessed by either hierarchical or relational search techniques. It is necessary
to code text characters in order to store, display, and print the text messages
created. Several standards are used to code text files, among which are ASCII
and EBCDIC.

Eventually video was to be a requirement for inclusion into text and
picture files, but by then the problem was well known and a key consideration
for solution in order for multimedia to flourish. When it became clear that
graphics and imagery storage was necessary, from the viewpoint of efficiency
and ease of handling, in the same database and then in the same file as the text
information, the data organization problem had to change. In some cases,
current systems have been modified to accommodate imagery and audio files,
while in other cases, new data storage systems have been developed from
scratch.

There are 2 principal image data management techniques, these being
image compression and image indexing. The former technique provides a
means of containing the contents of the imagery in a much smaller volume than
with uncompressed imagery. The other principal data management technique
is that of image indexing. Image indexing is a means of identifying and
categorizing images by storing image sizes, formats, compression schemes,
dates of creation and modification, collection sources, etc. The index is useful
in providing information about each image captured. These techniques will be
discussed further in later chapters.

Graphical/pictorial data storage systems, on the other hand, are relatively new technologies. They begin with the storage formats used for pictorial data. Some of these include PICT, TIFF, and others developed mainly by Apple or third-party providers who originally catered to Apple applications. These image storage capabilities are formats that were created to allow images to be recognized by different software application packages. These facilities will be discussed later.

1.4   Issues Involved in Multimedia Network Management

1.4.1   Psychophysical Aspects of Multimedia

Multimedia events involve what physiologists would refer to as experiences that utilize both sensory and motor skills, in interacting with each other. In a technical setting, the sensory experience translates into voice, print, and/or video conveyed to the viewer, i.e., the multimedia experience. The motor experience is referred to as interaction, and involves overt motor actions on the part of the user to effect the desired outcome. The motor skills of the user are used to make selections, or possibly request information. The whole idea behind multimedia is the incorporation of these motor and sensory experiences into the sessions involved between parties, so that they may have true interactive interchanges. Technologies have been developed and are being used to facilitate the motor capabilities of the human to activate the features and functions desired. These motor extensions range from touch screens, track balls, and the familiar mouse to, eventually, brain wave and eye tracking control.

1.4.2   Vendors Involved

It is hard to imagine a multimedia network that does not rely heavily upon all of the major computer players, such as Apple, the PC manufacturers, e.g., Compaq, IBM, Dell, or whoever may emerge as a major vendor of such architectures, and the manufacturers of multimedia storage devices, such as HP, and DEC. As the processing speeds and immediate memory capacities of the workstations increase, multimedia applications will become increasingly sophisticated. In the meantime, compression techniques, processor speeds, frame sizes, and frame rates will perform in a less than optimum fashion. These technical issues are critical to effective network management, and the video, audio, and text processing software will determine the efficacy of these modes of operation. For example, in the Apple and PC environments, a video processing package known as *Quicktime* is used extensively. The multimedia network management system can use such information to key on operational activity that, in turn, can help assess monitoring and control questions of interest to the network manager.

1.4.3   Operational Restrictions

Depending upon the speed of the hardware processor, video software products, such as *Quicktime*, can handle frame rates up to 30 frames per second to replicate those of full-motion video clips. The size of the frames, in terms of pixel width and pixel height, determines the resolution of the presentation, and the frame rate determines the motion characteristics, or "jerkiness" of the resulting video presentation.   Such subtleties of operation may escape recognition or consideration in the process of ongoing network assessments.

Additionally, the conveyance of voice, print, video, and interaction may require as many different types of transport media as there are sensory and motor modalities involved. This is a potentially complicated problem for a network management system that has to handle many different multimedia sessions simultaneously. Other transmission modalities could provide a single source of transport capability.

All the usable media capabilities could be carried over broadband transport systems, such as ATM or frame relay. ATM, for instance, is an example of a network layer service in the network management model, and fiber optics is an example of a physical layer service in that same model. The voice component of the multimedia system, if carried separately, requires a narrow band capacity, such as that found in copper. The text, or data stream, could also be transported via a copper system. Moreover, the present capabilities of narrow-band transport systems would allow both modes to be carried over copper or other narrow-band facilities. The various varieties of, what is referred to as broadband transmission systems, have a series of common threads, such as multiplexed data streams, and packetized compartmentalization of the modalities involved.

Imagery, and especially video frame sets, will require additional bandwidth capabilities. The update and refresh rates for imagery and the scan rates of video capture, from slow scan to full motion, will dictate exactly how much additional bandwidth is required for the picture presentation services planned. Sometimes, trade-offs are made between image quality and the number of transmitted frames per second. Image quality is an all-encompassing concept that includes filtering, pixel encoding, image encoding, and modulation techniques. New modulation techniques and compression schemes have assisted in supporting these needs, and these technologies will be discussed further in later chapters.

1.4.4   Multimedia Quality

The basic paradigm for data network management is as follows. An operator, sitting in front of an IBM 3270, presses the "Enter" key and nothing happens. The lack of response to such a keystroke may be a result of any number of problems beginning with the keyboard, and following through to the transmission system and, eventually, the data server.

The questions are where the problem is and what the nature of the problem is. The answer to the question of where the problem is may reveal a hardware failure and, therefore, answer the second question of what the problem is. If the problem is a software failure, locating where may be a more difficult issue. In that case, what the problem is may be a matter of either a software malfunction or a hardware lockup caused by the software, or possibly both.

The paradigm for multimedia is very different. The basic paradigm here is that the operator receives a poor-quality image, and the principal question is whether the received image is the result of poor origination or was corrupted during transmission and handling between the sender and receiver. This paradigm is illustrated in Figure 1-5. In the first frame, the image shown is the one intended to be sent to the receiver. Instead, the second frame was the one received.

The question, aside from the fact that the image received is unacceptable, is whether the corrupted image was the fault of the origination point or whether it is the result of a problem in the network. The first image may be corrupted by transmission errors or dropouts that could result in losses within and across each horizontal display line. Thus, as each packet of pixels is received in error or not received at all, its resulting affect could be a consistent white or black line fragment (pixel values 0 or 255) corresponding to that portion of the display line that experienced the pixel data dropout. Streaks in the frame imagery, therefore, may be the result of either poor imagery or transmission losses, or both.

These set of circumstances imply that the network and the management of this network which supports multimedia requirements will be as complex as the experience itself. For simple, standalone multimedia situations, such as for the access of sound and text-supported encyclopedia imagery, the network is integrated into the hardware in the form of the control hardware, e.g., mouse, the CD player, data bus, and the output devices, e.g., display. Thus, problems in that form result in either total outages or a very narrow range of choices. For more complex applications, such as those provided over LANs, the source materials are likely to be provided from servers or from repositories in specific workstations.

These systems are more complex in terms of their networking, interactivity between parties, and the overlying software needed to support such an arrangement. An open-ended, public-access multimedia system is even more complex, because of the possible sources and media types that could be utilized within the network, the variability of the access tools, and the size and variety of the physical means of providing the network.

In ensuing chapters we will see the technologies that attempt to harness these various modalities, and the architectures that manage these multimedia complexes. Some of these, especially those that attempt to squeeze added bandwidth from copper, attempt to operate at the limits of physics. Others, such as wireless technologies, rely upon compression to a great extent in order to conserve bandwidth; while others, such as fiber, require external power supplies, and a heavy investment in terrestrial installations in order to make their capabilities available to many parties.

**Figure 1-5
Multimedia Paradigm**

Intended Image Transfer

Actual Image Transfer

1.4.5   Multimedia Network Management Paradigm

The underlying utility of the process of network management is realized and implemented by software where functionalities allow users to share in conversations with each other, using databases and live interactive exchanges to affect the conversation. Users can even exercise abstract ideas to make points, or elicit comments through animation or the creation of synthesized sound. In essence, the users must be allowed to think, communicate, discuss, reference, analyze, create, and document a range of language needs. For example, when humans are together for the purpose of sharing views and making decisions, it is important that the interchange be immediate and accommodate several persons at once. An example follows.

User A may invoke a piece of text and graphic artwork that appears before all parties at the same time, similar to that facility created by software products such as *Timbuktu*. User B may begin to edit his/her displayed copy, while all users view these edited comments as they are being made. The progress of the ongoing conversation is available to all parties simultaneously. User C remembers a piece of artwork that works better with the text than that displayed, and pulls it up for simultaneous display to the other conference members. User A agrees and that graphic object is inserted into the slot previously reserved for the original art object. The overlying conversation about editorial changes, graphics replacements, etc., allows for instantaneous exchange of ideas, and decision making among the participants.

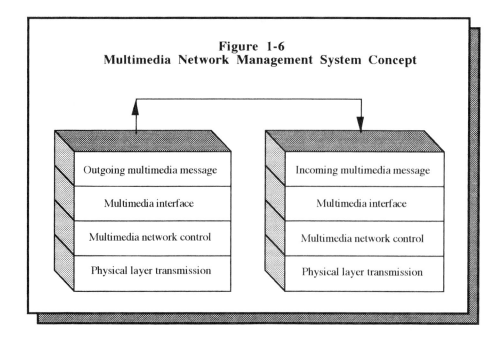

Figure 1-6
Multimedia Network Management System Concept

Outgoing multimedia message	Incoming multimedia message
Multimedia interface	Multimedia interface
Multimedia network control	Multimedia network control
Physical layer transmission	Physical layer transmission

While not absolutely required, multimedia usually implies the use of interaction between parties to achieve the desired results. This means that the same level of sophistication in network management must be achieved for all modalities involved, because of their interrelationship. And since multimedia is essentially a conversation between 2 or more parties, the same application must accommodate 2-way interactive communication. These and other requirements can be best understood by studying the established OSI model network management. The software functional concept of this condition is illustrated in Figure 1-6.

The basic OSI network management model defines 7 layers, from the physical layer at level 1, to the application layer at level 7. In the model shown in Figure 1-6, each interaction is a message between the sender and the various recipients, thus, it is synonymous with the application layer. The multimedia interface involves the establishment of the conversation, its rules of presentation exchange, and the end-to-end connection, e.g., addressing scheme. This encompasses the OSI layers 6, 5, and 4. The multimedia network control is analogous to layer 3. Finally, the physical transmission is similar to layers 2 and 1. The receiving side is a mirror image of the sending side. This arrangement makes it quite straightforward to isolate and control problems, because such have well-defined points of demarcation.

1.5   System Considerations for Multimedia Implementations

The evolution of multimedia capabilities has progressed along the same lines as data communications transmission has progressed over the years. First, multimedia capabilities were originally implemented via the internal data bus of the standalone computer. Next, they migrated to LANs where several multimedia positions could share in common technologies or interactive communications capabilities. Soon, computers with large storage capacities, known as servers, came to service users with access, communications, and printer utilities for those on the networks. Presently, multimedia functions are being accommodated in the designs of long-haul communications networks, such as ATMs, so that an unlimited number of multimedia positions can be supported simultaneously. The next stage is likely to be one of more detailed implementation of how humans communicate so as to better accommodate their needs.

1.5.1   Hardware

At first, hardware and firmware techniques such as high-volume cache memory, RAM, CPU processing rates, and eventually compact disk (CD) technology provided the desktop user with such software capabilities as optical character recognition, text-to-speech conversion, voice recognition, and audio compression, to name a few items. Later, LANs were developed that could accommodate bandwidths suitable for most of the multimedia modalities simultaneously, and in 2 or more directions. These LANs were then expanded, encompassing wider and wider geographic areas, using such identifying names as wide area networks (WANs), metropolitan area networks (MANs), etc., and incorporated additional features resulting in names such as FDDI, frame relay, ATM, and SMDS. The essential feature of these new capabilities, however, is packetized information transfer.

These specially named technologies rely upon established features for their realization such as data packet and compression techniques, and data multiplexing capabilities. The data packet approach provides for the digital representation of the information to be incorporated into a package of digital data that also includes the addressing and other housekeeping information, known as *overhead*. Each packet also includes indicators as to what each packet is and how it relates to the stream of like-type packets. *Multiplexing* is a method of combining several data streams together at the same time across several frequencies, or sequentially in time by dividing a channel into time slots for the allocation of data packets into an orderly arrangement. Thus, packet data and multiplexing are interrelated techniques and of importance to the successful deployment of multimedia systems.

Like most other technology developments, ideas have preceded laboratory developments and demonstrations, which, in turn, have led to the creation of prototypes and products. When it came time for prototype deployment, the communications availability was not yet a reality, but efforts soon began that pushed the bandwidth availability higher and higher for LANs, as well as switched and packet networks. Shortly afterwards, the deployment of rudimentary wide-band capabilities, such as ISDN, spurred wider-bandwidth developments, which in turn demanded newer developments in wider-bandwidth transmission capacity, now referred to as broadband transmission.

While it is true that wide-band transmission capabilities are far from ubiquitous, the possibilities for multimedia, at this point, are limited only by the human intellect. Human needs for communicating will define those possibilities for the future. The transmission systems needed to support these expanded interactive conversations, however, still have a lot of catching up to do, because certain transmission systems are marginal when supported by sophisticated data compression implementations. Data compression has become a major area for research and investigation for exactly this reason, and a development area for emerging audio, image, and video requirements. Compression techniques will likely remain a long-term requirement for both information storage and video presentations in the years to come. The reason is that data compression can be more easily pursued as a high-potential-payoff possibility for expanding bandwidth as opposed to research and development for new transmission methods, which has a high technology risk and high cost involvement.

Like the anatomy and physiology of human sensory and motor functions, multimedia reaches the user via divergent communications paths. Similarly, the user will dictate the nature and quantity of information for the intended recipient, which will utilize specific and non integrated transmission paths. Thus, voice, print, and video may likely be carried over different circuits, either physical or logical or both. It is likely that voice may be delivered to/from the multimedia desktop via familiar copper conduits, while print, i.e., text, will likely be carried over conditioned copper at megabit rates, leaving video to be conveyed via coax, fiber optic cable, or wireless means. Of course coax, fiber, etc., could support all of these modalities as well.

1.5.2   Multimedia Communications Concepts

The two-way nature of human communications means that all modalities will have to be bi-directional as well. Recent technology developments in integrated packet data transmission of various media, such as FDDI, frame relay, and ATM, may eventually support widespread multimedia capabilities, but only to metropolitan areas for the foreseeable future. It is fun to hypothesize about a world where an unlimited bandwidth is available via wireless or fiber means, but the reality is that current modalities may not soon be discarded.

We can begin to envision a conceptual view of what this means to a systems architect. Figures 1-7 A and B provide such a vision. In Figure 1-7 A, the multimedia integration workstation is essentially a computer platform equipped with the sophisticated software programs necessary to implement two-way conversations using voice, print, or video components. The user will have all modalities arriving and leaving at desired speeds consistent with the

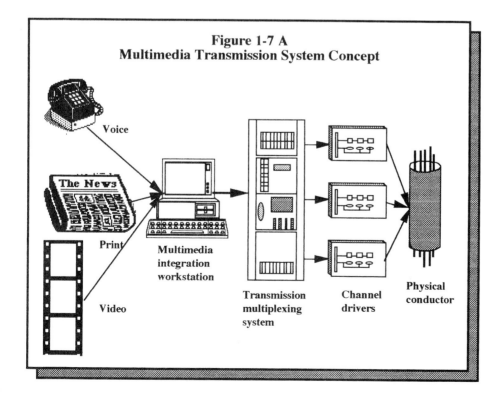

**Figure 1-7 A
Multimedia Transmission System Concept**

Voice

The News

Print

Multimedia
integration
workstation

Video

Transmission
multiplexing
system

Channel
drivers

Physical
conductor

data bus bandwidth available within the system, the transmission system, the decoding performed at the receiving end, and, of course, the absorption rates of the human audience. The workstation is attached to integrating transmission equipment that may employ packet systems to oversee the multimedia operations.

In Figure 1-7 B, the receiving side of the multimedia system concept is portrayed, which is essentially a mirror image of the transmission side. The functional requirements of the multimedia workstation are outlined as follows.

(1)   The workstation must be capable of accepting and transmitting and receiving voice, print, and video information. The requirement is consistent with the normal mode of human communications needs.

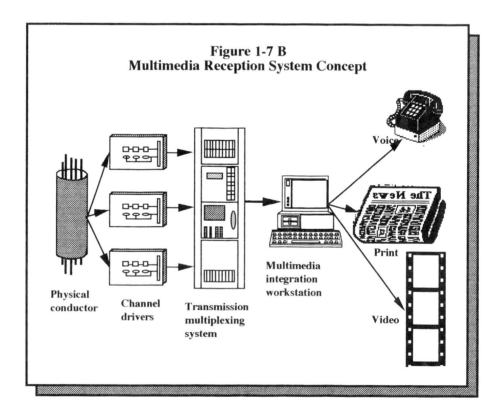

Figure 1-7 B
Multimedia Reception System Concept

(2)   The station must be capable of storing such information so as to provide an audit trail of the sessions. In some cases, the exact information must be stored; in other cases, summary information or references to databases may suffice.

(3)   The station must be able to access the same sources or repositories as the session partner(s), consistent with security regulations in force. Open or public information sources or databases should be available to both parties, or one party must be able to spool that data, as a file or message, over to the other member(s) of the session.

(4)   The station software must be capable of allowing simultaneous sharing of graphics notations, vocalizations, data accesses, key stroke actions, video clips, and the like. Capabilities must exist to allow manipulation of text, graphics, video, or animation by both or all parties to the session. The means that mouse, trackball, key strokes, or any other means of pointing must be communicated to and displayed by the receiving stations as they are being activated by the originator.

(5)   The station must be able to originate, edit, or receive source material so as to be a full participant in the media process.

Immediately after the integration workstation has had its chance to process the current conversation, the data are sent to the transmission multiplexor for packaging onto the physical medium. The multiplexing, drivers, and physical media are somewhat analogous to the lower 4 levels of the OSI network management model. Near the modal integration, the various modalities of voice, print, and video will be either split into separate channels for physical transmission to the destination point, e.g., copper, coax, or fiber, or they will be multiplexed into a common channel, similar to that function performed by the current ATM or Frame Relay technologies. At the receiving end, as depicted in Figure 1-7 B, the physical transmission system yields up the signal for conditioning, demultiplexing, and interpretation.

Thus, the multimedia system can only deal with the intricacies of 2 or more media if there is a complex system in place to effect such a solution. Future multimedia solutions will be more sophisticated and, hopefully, simpler in implementation.

1.6   A Look Ahead

The principal problems facing the typical network implementer who may need multimedia services include:

(1)   whether the applications services required for the creation and replaying of multimedia information are available and which ones

are needed for the realization of his/her network's operational
needs;

(2)   what network technologies are most needed and/or appropriate
        for the support of multimedia requirements; and,

(3)   what are the storage issues that will aid or detract from the
        handling of multimedia information.

The remainder of this treatment of multimedia networks and integration
will focus upon the analyses of these questions and their resolution. This will
be accomplished by examining specific topics in each of the chapters that
follow. The main focus is to be the network problems and their management,
the technologies that solve these problems, and architectures that accomplish
this objective. However, multimedia topics are introduced and discussed as
required to support the network management questions involved.
        The next chapter will begin to delve into the technologies associated
with the various aspects of multimedia transmission, data handling, and
applications services.

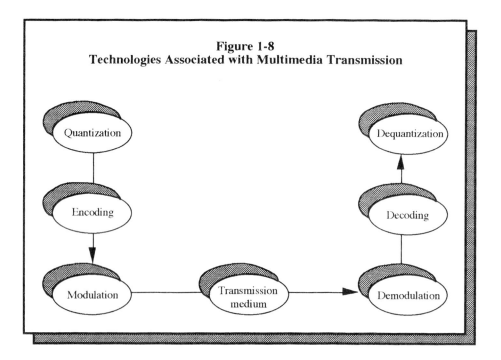

Figure 1-8
Technologies Associated with Multimedia Transmission

The technologies involved with multimedia transmission are illustrated in
Figure 1-8. First, the data or data requests must be quantized, or converted, in
some way so that the data can be readied for relay. Additionally, it must be

made compatible with the transmission system or to support the maximum range possible for the data being relayed to its destination. Second, the quantized data must be encoded so as to be transferred as efficiently as possible. And third, the encoded data will have to be modulated to be compatible with the transmission medium, whether it be light, radio frequency, or electrical.

Summary

The performance quality of multimedia systems is determined by the applications programs more than the network elements associated with the transmission of the communications traffic. These programs, and the originating information with which they deal, generate a certain level of result that may or may not affect the user perception of overall network quality. Thus, the discussion of multimedia networks and their management requires that the reader become aware of and appreciate the underlying technologies as well as various approaches to the packaging of multimedia products and services. Beyond this, the reader must also become aware of implementation architectures and applications issues. This first chapter sets the stage for the total discussion of these issues as they apply to an integrated approach to multimedia network management.

References

(1)   Avedon, D., and J. R. Levy, *Electronic Imaging Systems*, McGraw-Hill, New York, 1994.

(2)   Goyal, S. K., and R. W. Worrest, "Expert System Applications to Network Management," in *Expert System Applications to Telecommunications*, J. Liebowitz, (ed.), John Wiley, New York, 1988, pp. 10-50.

(3)   Harris, C. J., and I. White (eds.), *Advances in Command, Control and Communication Systems*, Peter Peregrinus, London, 1987.

(4)   Henning, W., "Bus Systems," *Sensors and Actuators A*, 25-27 (1991), pp. 109-113.

(5)   Hoffman, M., "Technology Profile Neural Networks," *Techmonitoring*, SRI International, July 1991.

(6)   Hsing, R., J. W. Lechleider, and D.L. Waring., "HDSL and ADSL: Giving New Life to Copper," in *Bellcore Exchange*, Livingston, NJ, March/April 1992.

(7)   Lemmon, A., *Marvel - A Knowledge-Based Planning System*, Internal Report, GTE Laboratories, Waltham, Mass, 1986.

(8)   Martin, J., *Design and Strategy for Distributed Data Processing*, Prentice-Hall, Englewood Cliffs, NJ, 1981.

(9)   McGrew, P.C., *On-Line Text Management: Hypertext and Other Techniques*, Intertext Publications, McGraw-Hill Book Company, New York, 1989, pp. 40-50.

(10)   Yager, Tom, *The Multimedia Production Handbook for the PC, Macintosh, and Amiga*, Harcourt Brace, New York, 1993, pp. 44-70, 129-206, 271-302.

Chapter

2

Multimedia Technologies: Audio, Text, and Images

Chapter Highlights:

Multimedia technologies that support two-way conversations are being used in such applications areas as storage media, transmission media, integration systems, and retrieval capabilities. These supporting technologies include data compression, bandwidth manipulation, and privacy control. New storage devices to be used for multimedia systems include compact disks, interactive, rewriteable laser disks, video jukeboxes, etc., in addition to the older devices, such as the CD-ROM, and standard video tape. The devices that are suitable for computer-based manipulation are those devices that are interactive and use digital handling techniques for the information involved. Thus, these 2 criteria will play a big role in the selection of multimedia devices for future use. The most realistic conclusion at this point is that the current trend in multimedia platforms will evolve into units that not only deal with multimedia conversations, but also create the data and storage repositories from which they come. In other words, video clips, text, graphics, and pictures can be created and molded into a desired multimedia presentation. This chapter will survey the technologies and techniques that encompass the audio, text, and image modalities of multimedia.

2.1   The Technologies of Multimedia Systems

There are certain basic technologies that support multimedia applications, such as data and image compression, transmission systems that maximize bandwidth capacity, packet switching and data handling systems, etc. These same technologies are also those that support any other application suite, such as the automated teller machine (ATM) for banking networks, or video teleconferencing for business applications, mainly because multimedia is actually an integration of some or all of the individual modalities involved in many other applications. For instance, cable TV is a service that carries video, data, and audio in one direction and, in some circumstances, data, in the form of signaling, in the other. Also, some data systems carry text, graphic, and image information bi-directionally, just as interactive systems carry information and signaling bi-directionally.

ATM systems are a good case in point when considering the key technologies for multimedia. ATM is a technology that, simultaneously, carries different media over the same transmission system by adding its own header to each packet as it is transmitted. ATM can accommodate packet video, packet audio, and a host of other packet data service protocols. The received packets are segmented and examined for destination addresses. Those that must be forwarded are again assigned headers and retransmitted, while those that are destined for local destinations are relayed to those outlets and further processed by their own special protocol decoders, e.g., X.25 systems. The ATM systems are similar to other multiplexer configurations in that they form their own network of linkages between local units, at which points ATM headers are added and striped as packets are transmitted and received. Thus, each of the mechanisms involved in ATM is used individually in its own networking scenarios.

The differences between multimedia and other modes of communication, such as data communications, are that:

(1)   Multimedia is an integration of 2 or more of the media modalities involved in its transmission and presentation;

(2)   Multimedia is a quality-related as well as a connectivity and integrity issue, and therefore requires the monitoring of its applications outputs in order to provide the network management system with the measures of the operational integrity of the multimedia system; and

(3)   Multimedia network management emphasizes both the physical and applications-level management of the operational systems interface (OSI) model of network management.

There is a whole compendium of technologies that can serve the needs of real-time multimedia programming, or the non-real-time generation of applications production [1:20-25]. The principal technologies involved in multimedia productions encompass such capabilities as the following.

(1)   On-Line, Magnetic Storage Devices -
These include the familiar diskette, hard disk, and other such devices. Permanent and temporary storage is consumed rapidly as the duration, format, and complexity of the application increases. Thus, data compression and adequate bandwidth capabilities are critical to the effective presentation of the desired information content. The biggest advantages of the magnetic storage technologies are that they are very cost-effective, have high capacity, and are very reliable.

(2)   Rewriteable Laser Disks -
Some technologies have been developed to allow the updating of laser disks as well as the ability to read and convey information previously written upon them. Rewriteable laser disk technology will be cost-competitive at some point and its utility will then be fully realized. The write once-read many (WORM) laser technology is based upon the use of ablative or burning means to designate spots on the laser disk surface that convey the 1s and 0s of the binary coding scheme. Technical Note 2-1 illustrates the technique used to implement the WORM approach for laser disks.

(3)   Compact Disk Read-Only Memory (CD-ROM)
Read-Only Disks -
Certain storage media are available in rewriteable forms, such as magnetic hard disks, while certain media only allow information stored on them to be read but not updated. CD-ROMs are among the latter type. Their utility is in their high density and total storage capacity, while their creation becomes cost-justified only if multiple copies of the same information is to be made available to the consumer.

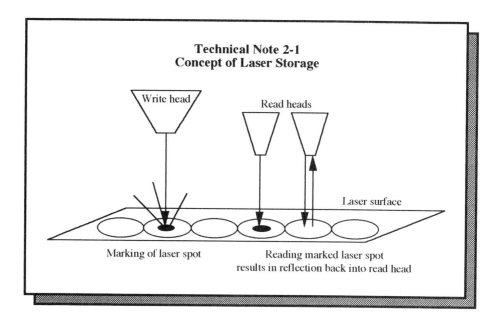

Technical Note 2-1
Concept of Laser Storage

Write head

Read heads

Laser surface

Marking of laser spot

Reading marked laser spot
results in reflection back into read head

(4)   Video Jukeboxes -
This is a technology inspired by the commercial record players of
the 1930s and 40s. Such devices had multiple records that could be
selected and positioned for play upon demand. The disk storage
equivalent has many platters available for positioning and reading
upon request. Such systems, therefore, have the ability to store
vast amounts of information. These devices have been targeted,
among other applications, for the playback of digital video
programs, e.g., movies [2]. Even a typical digital movie may require
upwards of 2 gigabytes (GB) of data. As a result, even a video
jukebox will need data compression capabilities in order to be
effective. The capacities of video jukeboxes are in the terabyte
(TB) ranges, but the number of video clips that may be requested
requires that such capacities be available.

(5)   Data Coding -
Many schemes have been devised to compact the data conveyed
over a given transmission line [3]. These techniques attempt to code
or to take advantage of the placement of information on the
transmission waveform so that increased segmentation of the wave
shape allows more and more information to be transmitted. This
technique was pioneered during the early days of digital
technology development by employing what was dubbed

quadrature phase shift key (QPSK). This will be discussed in more detail later.

(6)   **Narrow-Band and Wide-Band Transmission Systems -**
Narrow-band transmission systems include twisted copper pairs and even coaxial cable. These physical transmission technologies have been the object of intensive research during recent years so as to improve their bandwidth characteristics by coding and other schemes. Wide-bandwidth transmission systems include packet technologies, certain wireless systems, and fiber optics systems. These capabilities have the capacity for many interactive multimedia conversations simultaneously.

(7)   **Data Compression -**
Many software algorithms have been developed to increase the bandwidth of video and other multimedia offerings by compressing the information by mathematical or algorithmic manipulations. These techniques are simple in concept but sometimes complex in implementation. They revolve around the fact that data transmissions contain many instances of redundancy or repetitive occurrences.

(8)   **Animation -**
Animation has made tremendous strides in terms of presenting pictorial information in a way that is easy for the viewer to digest even though it may be inconsistent with reality. The typical examples of animation involve cartoon animals that can stay suspended in air, survive boulder hits to the body, and the like. Animation also provides a means of viewing processes from a standpoint that cannot ordinarily be appreciated during actual operations.

There are a number of other collection and presentation technologies that enhance the convenience of multimedia capabilities, but these are a step removed from the primary applications, storage, and transmission technologies identified above. Among these secondary technologies are charge coupled diodes (CCDs) used for image collection, and flat panel displays used for presentation. Technologies such as CCDs and flat panel displays enhance and support multimedia by making its application much more accessible and mobile for the user.

In this first major subsection about the technologies and tools of multimedia, a number of these products and services will be discussed that will, in turn, be integrated in ensuing chapters. We will first start with discussions of on-line storage capabilities and their evolution. Following this we will explore the newer technologies, such as the CD-ROM, write-once, read-many (WORM), and the laser disks. Later in this chapter, we will explore the limitations of the various coding and compression techniques, concluding with a detailed look at animation as one of the principal features of animation. This special treatment of animation is provided because animation provides views into the activities of living objects or directed activities of inanimate objects that could not otherwise be provided in real-life photography or drafting.

2.2   On-Line Storage Technologies

On-line, or peripheral, storage is primarily provided through the employment of magnetic disk technologies that utilize one or more circular platters to store the data of interest [5:45]. These storage devices consist of one or more round magnetic surfaces that can rotate at very high rates of several thousand revolutions per minute (RPM), and can be stacked and ganged together for greater data handling capacity. Stacked platters provide greater storage capacity than single platters and shorter seek times. Storage efficiencies and storage speeds are enhanced by defining cylinders within platter groupings, where a specific track across a group of platters is defined as a cylinder. This cylinder can be quickly located by the set of read/write heads servicing the array of platters involved, and information written and/or retrieved can be more easily and quickly transferred.

Each platter is divided into a number of tracks, each running concentricly relative to the others that run around each platter at different radii from the center of the platter. Similar tracks in a number of stacked platters define what is called a *cylinder*. A cylinder is an imaginary cylinder passing through 2 or more platters at some specific track location. The tracks in each platter are, in turn, divided, by means of software partitioning, into sectors, and each sector is divided into a number of the basic disk storage units, known as *blocks*. The database and data management aspects of these storage technologies will be discussed in more detail in Chapter 4.

2.2.1 Compact Disk Technology

Compact disks (CDs) used for computer data storage and retrieval are identical to those CDs used for audio playback. The difference is what is done with the data as it is converted from digital to analog form prior to its output, and where the data is directed, i.e., either to audio speakers or to video monitors. Recording, erasing, and re recording of CDs is not widely available to the average user of such media. Their operation is restricted with present-day cost-effective technologies to a read-only mode of operation. However, it may be only a matter of time before the consumer has such features as recording and overwriting available for everyday use. Thus, the present-day CD technology available to the consumer market is read-only for the most part. The familiar 4 $3/4$-inch-diameter CDs store 600 Mb, which is roughly equivalent to the capacity required for storing over 250 novels in digital format. CD players allow the user access to interactive services via the software control associated with such units. This capability is allowed by the fact that the system is operating via the internal computer communications bus for the most part, as opposed to reliance upon standard communications facilities. For distances greater than those internal to computers, new physical and functional technologies must be implemented in order to obtain the desired interactive features of CDs. These new technologies rely upon compression and high-speed transmission hardware utilizing packet or circuit switching facilities.

2.2.2 Magnetic Storage Technology

This subject will be dealt with in detail in Chapter 3, however, it can be said that magnetic storage is the best available technology at this time because it provides high capacity, low cost, and reasonable retrieval speeds, as compared to other possible choices for the same facilities. For example, magnetic storage can provide more than a gigabyte of storage for about $4.00 per megabyte. Because magnetic disk storage technologies have been used so much over the past several years, they are now highly reliable. Instances of disk failures are very rare. In addition, typical latencies or delay times between a data request and data retrieval have decreased from 30 ms or more to around 15 ms.

2.3   Image Collection Technologies

2.3.1 Document Scanners

The document scanner is based upon technology developed for military satellite programs. The basic component of the scanner is the *charge-coupled diode* (CCD). This is a light-receiving element, which together with a large number of other such elements arranged in a row, provides a matrix to represent the input on a line-by-line basis. The definition of a line is somewhat variable, but it is roughly a swath that is a fraction of a millimeter long and wide enough to accommodate the width of the document being scanned. The basic idea is for the document being scanned to be illuminated, the reflection from which is collected by the CCD element array. These collected values are converted to digital format and stored while the next CCD line is scanned, and so forth. The concept is for the CCD to pass a higher voltage for white areas and a lower relative voltage for darker areas.

Some other technologies have been implemented in place of the CCD, such as the laser and the light-emitting diode (LED). Scanners sometimes come with, or have available to them, optical character recognition (OCR) software that will take the scanning process to the next step, which is to interpret the image in terms of the characters or graphics represented on the page. Scanners operate at various resolution densities, ranging from 72 dots per inch (DPI) to 600 or more DPI. For a 200-DPI image scan of an 8 $^1/_2$ by 11-inch page the scan time is about 2 to 3 s, after which the document is temporarily held in storage and then permanently stored away. Sometimes image compression is performed upon the image at this point so that the image storage needs are minimized.

One of the big advantages and values of multimedia systems and their accompanying networks lies in their ability to receive data from various sources, and to output that data, or modifications of it, to many different receivers. Document scanners satisfy one of these important input requirements based upon the nature of the requested input. Devices that scan information and store it in computer file storage formats are referred to as document scanners. The storage of scanned information can either be in image format or a combination of text and graphics formats depending upon whether the incoming data has been subjected to optical character recognition (OCR).

These devices reflect light cast onto the image surface back to sensors capable of resolving very small reflection areas. These reflection areas determine the minimum resolution of the system, and the number of points per area is given in dots per inch. These reflections are then stored as values representing varying degrees of gray or shade of color. Scanners provide the capability to capture images, graphic, text, or both, that are otherwise not available from such sources as CDs, disks, tape, etc.

Scanners capture images at differing levels of pixel density, e.g., 70 DPI, 300 DPI, 600 DPI, etc. These density levels determine the resolution of the image captured to a certain extent. For instance, if an image is captured at 70 DPI, then its resolution is fixed at that level no matter what its capture density is. Even if the scan rate is 300 DPI, the resolution is restricted to 70 DPI. On the other hand, if the source image is 600 DPI, then the scanning rate can determine the captured resolution up to 600 DPI, based upon the value selected for scanning. Typical document scanners can perform scanning at 300 DPI, which is considered medium-quality resolution. Some professional document capture systems can collect images at resolutions of 1000 DPI or more.

The data scanned into the computer is initially stored as graphics data, like a picture or line drawing. If text is part of the image scanned, the text becomes part of the graphics image itself. Thus, the text cannot be manipulated with word processors, because the characters are not yet in a character standard format, such as ASCII. After capture, the document scanners rely on various compression algorithms that facilitate the volume of data storage required.

An adjunct capability to document scanners is their (sometimes) ability to interpret characters within the documents scanned. This capability is called optical character recognition (OCR). Because of its importance in document interpretation, OCR will be discussed in a separate section to follow.

2.3.2 Optical Character Recognition

The interpretation of text within an image is an important consideration for document usage. There are millions of documents that were originally written prior to the advent of the word processor. Reading and appreciating these documents from CD-ROM or other media can be performed without the aid of document capture and/or OCR. However, if editing or cropping of any of these documents is needed, either the text within the captured image must be broken out and converted to standard text codes understandable to computer word processing systems, or the text must be reproduced by retyping.

In order to get around problem of retyping, which would be prohibitive financially, optical character recognition is the option of choice to be performed upon the image containing the text. OCR is software-based and involves the ability of the software to identify various lines and splines as being characters belonging to some font type and point size. The OCR software resolves all the characters that it is able to resolve, and designates all other lines and splines as part of the graphics region.

The original image is scanned so as to result in what is referred to as a *bit map*. This bit map replicates the text characters in an essentially pictorial form. An OCR system reads the bit-mapped images, and interprets them as ASCII characters. Obviously, the system is not flawless, due to different circumstances. For instance, the scanned document may be dirty because of the transmission system or the scanner itself. Possibly, the text may have been scanned on a slant, or the original document may have been in color. Whatever the situation, these flaws may prevent total accuracy in the OCR process, so the operator, on occasion, will have to manually enter the correct ASCII character that represents the text letter in question. Most OCR systems have the capability to be trained so that they can begin to recognize even difficult or noisy character representations. Even OCR training, however, cannot achieve 100% accuracy for OCR, but this feature does help correct the easier scanning problems.

2.3.3 Frame Grabbing

Several software packages, along with the necessary hardware board, are available that have the ability to capture frames in a full-motion video stream. These programs operate by being able to detect synchronization signals at the start of each frame and to collect the video until the start of the next frame. The program can also send the frame to a digital conversion (sample and hold) circuit that digitizes the video signal to a desired level, such as 8-bit, 16-bit, etc., level of depth.

2.3.4 FAX Capability

FAX is yet another technology that can be manifested as either a standalone hardware device or as a software capability within a different device, such as a computer. This form of communications has been regulated for some years by the International Standards Organization (ISO). Different versions of these FAX standards have been released under the names of Group II, Group III, etc. The Group standards specify such maximum data rates as 4800 and 9600 bits per second (bps). In addition, the FAX standards also specify scanning parameters.

The scanning process operates a line at a time. The clarity, or resolution of a FAX system depends upon the density of the scanning activity. The normal rate is 100 scan lines per inch. Thus, a standard $8 \, ^1/_2$ by 11 page requires about 850 scans. Each scanned line contains approximately 1730 pixels, or picture elements. Thus, the entire page contains over 1 million bits of information. At 4800 bps, with no compression, requires about 3.5 min. Such unacceptable transmission rates require that compression be employed so that transmission times can be reasonable.

2.4   Image Processing

Visual information can be addressed as either still images or video. Video is a set of closely sequenced still images that give the illusion of motion. This section will address only still image processing. The image collection and processing system is illustrated conceptually in Figure 2-1. The image has to have some means of providing contrast or texture in order for photons of light to be reflected from a source back to the receptors of the viewer, or collector system.

Even so-called low-light image collection systems, such as those used by the U.S. Army, must have some degree of light reflection in order to be useful. As the collector receives the photons of light or after the entire image has been collected, the pixels of the image are digitized according to color intensities, or shades of gray. In order for images to appear reasonably smooth to the human eye, a minimum range of over 100 intensity levels is usually required [4]. The image pixels are then stored on-the-fly as the image is being digitized. Once the entire image is collected and digitized, it may be subjected to image enhancement algorithms, after which it can be compressed for such reasons as storage efficiency, transmission, or basic handling.

The capture, holding, and handling of imagery begins with the sampling of the raw image. This step results in the storage of the image in temporary storage or a buffer area of memory. Following this, the image is subjected to the image enhancement process. There are several methods employed to increase the resolution or detail of an image, which will be discussed below. After the image has been cleaned up, it may be compressed or sent to a more permanent storage area in memory. Image processing is now used in such commercial or military enterprises as medicine, for diagnostic medical imaging, in satellite image processing, and in TV or news image transmission.

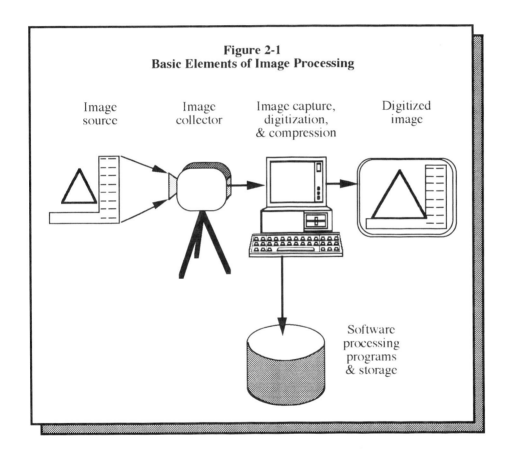

Figure 2-1
Basic Elements of Image Processing

Since the early days of image processing, tremendous advances have been made in the field of image storage, manipulation, and enhancement, such that otherwise unintelligible images can often now be made intelligible. The biggest advance has been in the area of pixel technology as opposed to character-based image creation. Some of the more important methods used in image enhancement processing are described below. A brief background to image processing is presented in Technical Note 2-2.

Technical Note 2-2
Image Processing Background

Image processing was initially developed as a result of the need to improve the intelligibility of trans-Atlantic digital images sent between New York and London via submarine cable [4]. These communications systems were originally laid in the late 1800s, and image transmission time across the Atlantic was lengthy. In the 1920s, the Bartlane cable transmission system enhanced both the quality and transit time of images across the Atlantic from a week or more to less than 3 hours. The system worked by encoding the signal for cable transmission into a series of pulses, similar to Morse code, and then decoding the signal at the receiving end, after which the digital codes were printed on specially designed printing equipment. The essential equipment used was a modified telegraph printer with special typefaces such that each code caused a typeface to be printed in sequence across the page, a line feed was activated, and the next line printed one character at a time. The printing process included characters, such as Xs, Os, #s, *s, or other suitable characters that would convey the impression of the correct strokes used in the creation of a picture in the mind's eye. This type of image processing is illustrated in Figure 2-1.

The perception of brightness was originally conveyed in the Bartlane system using 5 different characters. Later, the system was expanded to provide 15 gray levels, or shades of darkness. This type of image conveyance and transcription approach did not change appreciably until 1964, when the world of image processing took on renewed importance with the advent of space probes and satellite photography. In 1964, the Jet Propulsion Laboratory (JPL) in Pasadena, California, processed Ranger 7 moon probe pictures that were radioed back to JPL in digital format. Each pixel was represented by a series of bits that conveyed the brightness level of each pixel in the image. Not only did JPL computers reconstruct the pixels from the digital words recovered, they also performed enhancement of the imagery. By subjecting the digital words to analysis, the programs were able to determine when certain pixels were returned in error, and further to replace these pixel values with the proper ones.

Bartlane Cable Transmission System

Bartlane code system for image transmission. Three sides of a cube appear as different ASCII characters. When combined they take on the appearance of the desired image.

2.4.1 Image Enhancement Using Histogram Modification Techniques

Computer images, similar to those experienced with FAX transmissions, often result in outputs that lack optimum contrast, that are smeared, or that are otherwise corrupted with noise in the form of extraneous marks on the page or display screen. Such images can often be subjected to a technique referred to as *equalization*, where all pixels are modified either by adding to the value of each or subtracting some value from each.

This technique is implemented in the following way. Develop a probability density function of all pixels in the image. Next, generate a transformation function for the probability function. This is accomplished by integrating the probability function over the interval of its existence. The old pixel values are now converted to their new values by finding their original values on the transformation function, and then generating their new values. The results can be quite spectacular. Figure 2-2 shows a comparison of just such a transformation as it is applied to an image.

The general idea is to decrease high-valued pixels (bright spots), and to increase low-valued pixels (dark spots) such that there is an overall normalization of the total set of pixel values. An example of how the technique works is illustrated in Figure 2-3 and is explained as follows.

The probability density function of the original image is developed. The function is then integrated to produce the transfer function for conversion of the original pixel values to the new values. This transfer function, when applied, will adjust pixel values to ranges that either increase or decrease the image intensity.

2.4.2 Image Enhancement Using Image Smoothing Techniques

This technique removes spurious noise effects that may be present in the digital image. These spurious effects are usually the result of poor sampling, low-quality storage media, or corrupted transmission channels. Image smoothing can be realized by either spatial or frequency techniques. A discussion of the various types of frequency filters that satisfy image smoothing requirements is presented in the next section. The first of these, the spatial smoothing technique, is implemented by using a neighborhood averaging approach. This means that pixels surrounding the pixel being averaged are used to determine the value of the pixel in question. This technique may be implemented by application of the following mathematical formula:

$$g(x,y) \quad = \quad {}^{1}\!/_{M} \sum_{(n,m)\,\varepsilon\,S} f(n,m)$$

where

g(x,y)   =   the smoothed image space within the range of
an N by N image

Figure 2-2
Image Enhancement Using Normalization Technique

Abnormally Bright Image

Normalized Image

$$^1\!/_M \quad = \quad \text{the reciprocal of the total number of}$$

points within the image space

$$\sum_{(n,m)\,\epsilon\,S} f(n,m) \quad = \quad \text{the integration, or summation, of points about}$$

the point in question, defined in this
case by g(x,y), n and m being a subset
of the entire sample space, defined by S

Thus, a nearest-neighbor smoothing algorithm, where the radius about the point
to be smoothed is 1 pixel, is reduced to:

$$S \quad = \quad \{(x,y+1), (x,y-1), (x+1,y), (x-1,y)\}$$

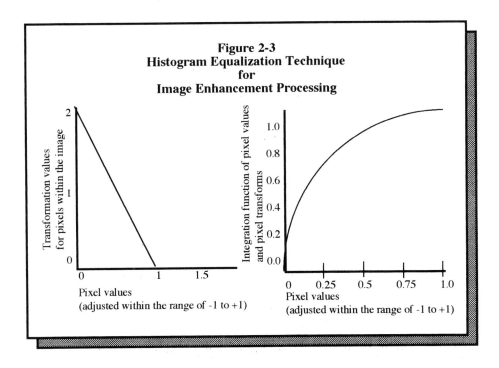

Figure 2-3
Histogram Equalization Technique
for
Image Enhancement Processing

Figure 2-4 shows a couple of approaches to the spatial approach. And, for the
smoothing radius = Δx, the above equation holds.

For a radius $= \sqrt{2}\,\Delta x$, the following equation holds:

$$S \quad = \quad \begin{aligned} &\{(x\text{-}1,\, y\text{+}1),\, (x,\, y\text{+}1),\, (x\text{+}1,\, y\text{+}1),\\ &(x\text{-}1,\, y),\, (x,\, y),\, (x\text{+}1,\, y),\\ &(x\text{-}1,\, y\text{-}1),\, (x,\, y\text{-}1),\, (x\text{+}1,\, y\text{-}1)\} \end{aligned}$$

As the value of Δx increases, i.e., the values in the nearest-neighbor pixels increase, the smoothing, or blurring, effect is exacerbated. The challenge, therefore, is to use as few neighboring pixels as necessary in order to dull the noise effect while maintaining the resolution of the image being processed. Another variation on this approach provides for thresholds to be used so as to negate the blurring by first comparing averaged pixel values to those values of their neighbors, and not applying the result unless the averaged pixel exceeds a certain threshold value. This is implemented using the following formula:

$$g(x,y) = {}^{1}/_{M} \sum_{(n,m)\,\varepsilon\,S} f(n,m) \text{ if } \left| f(x,y) - {}^{1}/_{M} \sum_{(n,m)\,\varepsilon\,S} f(n,m) \right| > T$$

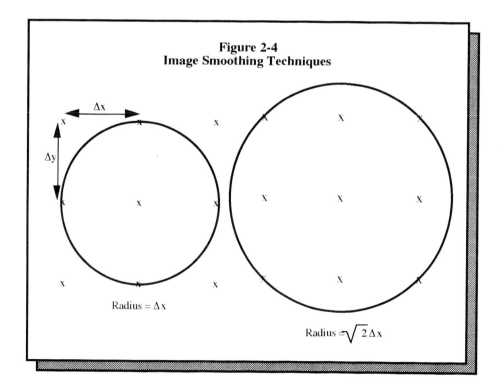

Figure 2-4
Image Smoothing Techniques

2.4.3 Image Enhancement Using Low-Pass Filtering Techniques

There are several types of frequency filters used to smooth images and eliminate, or at least control, noise corruption in the images. Figure 2-5 shows the concept of the low-pass filter. The ideal filter is one that has a sharp cutoff at a predefined frequency, D_0. Other low-pass mathematical frequency filters have more sloped aspects to their frequency roll-off characteristics.

$H(u,v)$ is the frequency transform that is to be applied to the original image frequency function, and u and v are the individual frequency components of that function. Thus, the function $H(u,v)$ has frequency components in the x and y directions. Its shape and influence upon image smoothing is determined by the slope of the function. There are mathematical versions of the low-pass frequency filter, such as the Butterworth filter, exponential filter, and the trapezoidal filter. When a low-pass filter is used, on the one hand, it helps eliminate noise, but, on the other hand, it blurs the image by the same action. Low-pass filtering, therefore, may be used in conjunction with another filter that will enhance edge detection. The generalized low-pass filter equation is:

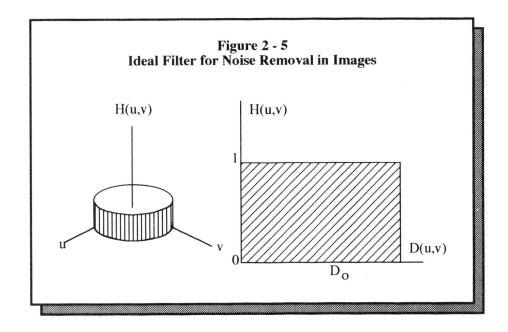

Figure 2 - 5
Ideal Filter for Noise Removal in Images

$G(u,v) = H(u,v)$ by $F(u,v)$

where
$F(u,v)$ =   original image transform

H(u,v)   =   low-frequency transform
G(u,v)   =   resultant image transform

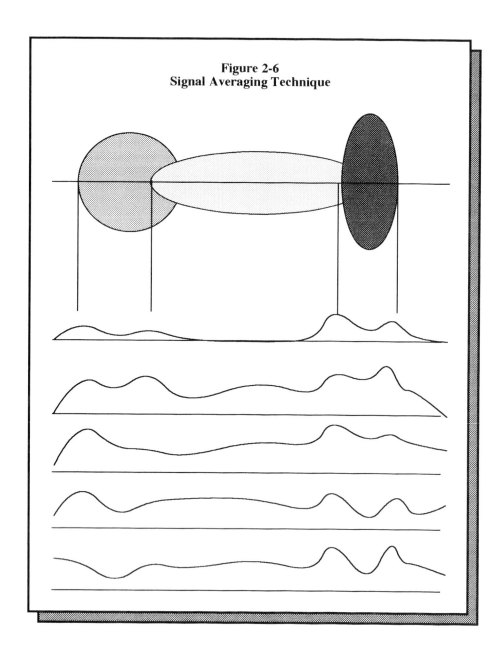

Figure 2-6
Signal Averaging Technique

2.4.4 Image Enhancement Using Signal Averaging Techniques

Signal averaging is a technique for lifting stabilized, or periodically recurring, images out of noisy environments. The theory behind signal averaging is that noise is assumed to be random in nature. On the other hand, a meaningful image has periodic contrast areas that represent edges in the image space. When images are overlaid with background noise, the image may look corrupted, or indiscriminate. Figure 2-6 shows a test image consisting of 3 overlapping ellipses. A line has been drawn through the image identifying one line of pixels.

Figure 2-7
Animation Representing a Skyline View

Below the image, that particular line of pixels has been turned so as to view it on edge with pixel values represented by the curve shown. According to the signal averaging theory, each line of pixels, as shown, can be subjected to random noise, as shown by the next 4 line representations. If the pixels in a large number of such noise-corrupted lines are added and then averaged, the noise will be averaged out of existence, while the image edge values and texture will be preserved. This fact can be somewhat appreciated by looking at the 4 noise-corrupted image lines. If these 4 lines are added and averaged, the result will yield something similar to the original line value.

2.5   Animation Facilities

Any treatment of multimedia techniques and tools has to include some discussion of the subject of animation. This capability, although rooted in the world of children's cartoons, has had a tremendous impact upon teaching and information conveyance in the more traditional adult subjects of finance, business, engineering, and science. The world of animation addresses not only characterizations of living objects, but also representations of real-world activities and simulations of textual information. All of these visual activities combined with movement and alteration can be enhanced with audio sounds that augment the natural or animated imagery.

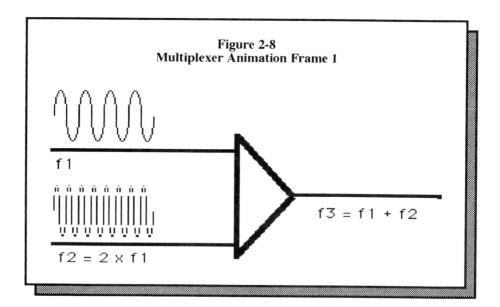

Figure 2-8
Multiplexer Animation Frame 1

f 1

f2 = 2 × f1

f3 = f1 + f2

Animation of imagery or cartoon-like objects can add emphasis, clarity, simulation, or time and spatial travel through a subject not ordinarily possible in the real world. This segment of the chapter will focus upon some of the features and functions that can be satisfied with animation tools. The importance of animation to the discussion of multimedia, and in particular the management of multimedia networks, is 2-fold. First, animation can be used as a benchmark for the evaluation of network performance, due to the fact that animation clips can be standardized, and that standard can be made known to management elements of the system for analyses of the system. Second, as mentioned earlier, the applications of multimedia sessions determine the network management required, and not so much the actual transmission components of the data. Therefore, system quality becomes a function of the applications that utilize the system, and animation can be made a known quantity for such evaluations.

Figure 2-7 shows an example of the use of animation as a tool for representing a real object in a way different from reality. In this view, the spacing between the prominent features of the San Francisco skyline is shown in a contracted form in order to get everything into the view. In reality, these objects are much further apart from any angle. The use of animation can be used to bring pieces of an object together, explode them for better viewing, or present them in a way that defies gravity, as an example.

Current and future trends in animation provide for an even further expansion of our imaginations by representing humans and other animate objects in ways that challenge us or facilitate our perceptions so as to expand our appreciation of these views as they move or as they are presented to us. Additionally, the viewing of landscapes or other scenes of interest can be made to add features or viewing angles that otherwise would be impossible or unlikely for us to see on a regular basis.

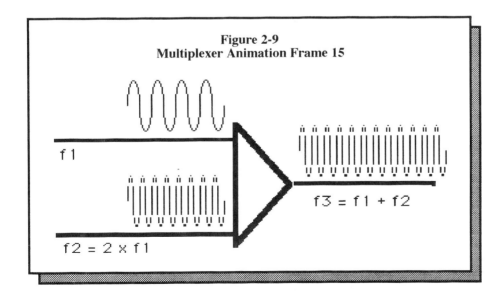

Figure 2-9
Multiplexer Animation Frame 15

f 1

f2 = 2 x f1

f3 = f1 + f2

Another example of the power of animation is its use for illustration. Shown in Figures 2-8 and 2-9 are 2 of many frames from a sequence that illustrates the concept of data multiplexing.

The first frame shows 2 signals traveling along their respective input paths toward a device that combines them to yield a different waveform of at least a frequency equal to the sum of the 2 input frequencies. Frame 15, shown in the second figure, shows the resultant waveform frequency that is the result of the addition of the 2 input frequencies.

2.6   Other Technologies

There are several miscellaneous technologies that have the potential to enhance the principal media components of the multimedia environment. These technologies, such as text-to-speech conversion, are direct linkages to the various media and, in the case of this particular one, provide a capability to convert between media types. Identified below are some of the auxiliary techniques that can be utilized to support multimedia applications.

2.6.1 Knowledge-Based Systems

Database, or knowledge base design for the support of expert systems inferencing, has been a legitimate pursuit in computer science for many years [5]. The major difference between a database and a knowledge base is that in a database the retrieved object is a fact, whereas in a knowledge base the retrieved object is a deduction. In addition, the knowledge base must understand queries and respond in natural language fashion. Multimedia systems may, in a short time, utilize the features and advantages of knowledge-based systems to create animation, interpret images, generate voice and text presentations, etc.

2.6.2 Theorem Proving

The expert system must be capable of proving an assertion, or theorem, e.g., that a particular fault x occurs at location y. This is somewhat akin in an opposite sense to the null hypothesis, where the assertion is that there is no difference between a particular event and its consequences. Theorem-proving techniques attempt to establish procedures for selecting among several rules to apply to the problem and to establish subproblems leading to the solution. A theorem usually has a certain level of uncertainty attached to it, so that the proof, in turn, has a certain level of probability attached. These techniques will be used in conjunction with hypermedia tools to prove or disprove assertions that are proposed in integrated document requests.

2.6.3 Automatic Programming

All compilers are really *automatic programming* tools, because they convert certain syntax, known as source code, into executable object code that the computer can actually operate upon to create the desired results. Such code consists of the list of instructions that can be directly read by the computer and executed as ordered. Automatic programming takes this concept another step further into the realm of automation. In automatic programming, high-level descriptions of need are themselves processed into executable object code. These high-level descriptions may be loosely structured language, known as pseudo-code, that is processed through a natural language processor, then interpreted, and finally subjected to an object code generator. These high-level representations of the programmer's requirements can be expressed either as text or as visual icons that can be maneuvered into a sequence for establishing the line of processing.

2.6.4 Solutions to Combinational and Scheduling Problems

The traveling salesperson problem has been one that has received considerable mathematical interest for many years. It is one in which the issue is how to schedule the rounds of a traveling salesperson so as to maximize his or her customer exposure over some period of time, or said another way, to minimize his or her travel time between customers. Solutions to these and related problems generate an explosion of possibilities. Typical solutions require an increasingly large time frame in which to resolve the near-optimum possibilities sought. AI approaches operate so as to constrain these solution possibilities by using domain information. Such domain-specific information might be appointments or other needs that would negate some of the possibilities. Schedule problems may soon be issues whose solutions are aided by multimedia techniques. For example, simulation coupled with visual aids can enhance the perception and impression of scheduling solutions. These combinations of visual aids, audio information, and the simulation processes will provide a new dimension to problem solutions as modulated by multimedia tools.

2.6.5 Machine Perception

There are two basic forms of machine perception, one being vision and the other being language or speech. The understanding of human spoken language by a machine is an example of speech perception, and vehicular road- or terrain-following perception is an example of machine vision. In the field of telecommunications, both types are of equal importance. Mapping and exploring, geopositioning, and object location are just a few of the possibilities for machine vision. With regard to speech perception, the receipt, decoding, and interpretation of language is of obvious interest to telecommunications. Many speech perception systems have already been developed, a large portion of which have been developed by the toy industry. Such systems will find their way into more widespread use in the near future.

2.6.6 Expert Consulting Systems

Expert consulting systems are oriented toward solving problems and many times can be an adjunct to an existing on-line expert system. These systems interact with the human expert to receive rules, or may carry on a dialog with the human operator to collect ancillary data, or might test specific hypotheses requested by the operator. One early such consultant was the SRI expert system known as Surveyor. This system was designed and implemented to identify possible geological locations for valuable minerals and ore deposits. It was given credit for actually locating a rich vein of valuable ore. Other expert consultants include Stanford University's MYCIN, used in the diagnoses of certain bacteria and diseases. Expert systems of all types are fertile areas for the use of various media to enhance, demonstrate, or support the intended purposes of the expert consulting system.

2.6.7 Robotics

Robotics is an exploding technology in the product and service delivery arena, and will impact telecommunications to a great extent in the next few years. Potential areas of usage are maintenance and repair, operations monitoring, and security. Robots are classified according to international standards, and have potential that is only partially understood at this point. The Japanese classify a device that is used to pick up an object and place it somewhere else as a robot. The Robot Institute of America as well as the British Robot Association classify a robot as *a reprogrammable device designed to both manipulate and transport parts, tools, or specialized manufacturing implements through variable programmed motions for the performance of specific manufacturing tasks.*

Robots are the last step into the realm of multimedia, because, almost by definition, they must be capable of receiving imagery, audio, and text information as well as being able to transmit such modalities. In other words, they must be capable of the same sensory and motor functions as a human. As a result, there are numerous opportunities for multimedia to assist robotics and related technologies. These include computer-aided design (CAD) and computer-integrated manufacturing (CIM). Each of these may soon be integrated with the various attributes of multimedia to enhance their capabilities, and robots will be used to further the uses of these technologies.

2.6.8 Natural Language Processing

Natural language processing is an important topic from a human interface point of view. The basic objective for this line of interest is to make the machine capable of understanding unstructured inputs from the human operator, whether they be spoken, typed, or pointed out. One of the best examples of natural language usage is the use of icons as used on the Apple or Microsoft human interface. The point, drag, and click approach is a natural extension of the human tendency to reach out and grasp the object of interest.

2.6.9 Computer-Aided Instruction

Learning is a process best enforced by a variety of methods, some of which are programmed instructional training, simulation, hands-on practice, classroom teaching, visualization through such media as video programs, and computer-aided instruction (CAI), to name a few. In this approach the instructor's expertise is captured and replayed to the student to be tested against the expert. There are varying degrees of effectiveness gained from such means, because the realism is more limited than the simulator, although less expensive.

2.6.10   Topological Reference

The latest approach to troubleshooting involves a schematic diagram that is read into the system, its symbols interpreted, its interconnections defined, and its operation is inferred. The existence of trouble is marked on the screen of the system depicting the schematic, the topology is examined, and inferences about the system's failure are made based upon the graphical picture of the device that was captured. The vision, therefore, is to "read in" a schematic and to diagnose faults based upon indicated errors within the domain of the graphical boundaries of that schematic.

2.6.11   Text-to-Speech Conversion

One of the more useful technologies that currently supports multimedia implementations is the text-to-speech technology. This capability is implemented by using a set of rules such that when letters or a group of letters in sequence are recognized, the corresponding sound is retrieved from memory and queued up for conversion to analog form and feed to a speaker.

2.6.12   Voice Recognition

Many approaches to the understanding of human vocalizations have been attempted. Recently, these attempts have begun to produce some reliable and consistent results. Voice is the ultimate point-and-click tool. As the technology becomes better and better, its utility as a software-driven device for use in the multimedia realm enlarges.

2.6.13   Tone Decoders

Interactive voice response (IVR) systems have been used with great success, in terms of the consistency and reliability of the technology, for several years. These and other audio text-style menu-driven systems have enhanced the usage of information transfer. Such capabilities are implemented on plug-in computer boards that are inserted into expansion slots inside the computer chassis and interpret incoming tones as numbers, or, alternatively, letters of the alphabet. These indications can then be stored into a database or used to activate responses back to the original sender.

2.7   Basic Components of Multimedia Services

Figure 2-10 provides the reader with a pictorial view of the key ingredients of the multimedia problem. Each of these will be discussed at various points in this text, or have already been introduced in this and the previous chapter. Basically, the media network consists of:

(1)   an access device that might range from a multimedia palm-top computer, or pad, to a desktop system complete with CD-ROM, high-capacity disk, and high-quality speakers;

(2)   a queuing and buffering device that strobes the data at the optimum rate consistent with the needs of the system's transmission, reception, and queuing constraints. For video and audio requirements, the data flow must be isochronous;

(3)   the transport network access equipment over which formats and controls provide access to the network. Such a transport system should be designed so as to minimize the data formatting overhead, and to make the data flow as isochronous as possible. Transport systems do not usually have sufficient bandwidth to convey the information of interest without compression [3];

(4)   the transport network that carries the multimedia information; and finally,

(5)   the multimedia database that serves as the repository for media information, such as video and audio, which must be strobed in and out at rates consistent with real-time rendering needs.

The multimedia database is one of the most critical aspects of the entire media network, because it has to be operated so as to accommodate the display of video at video rates, a condition known as isochronous information transfer. This problem will be explored in more detail elsewhere in this text.

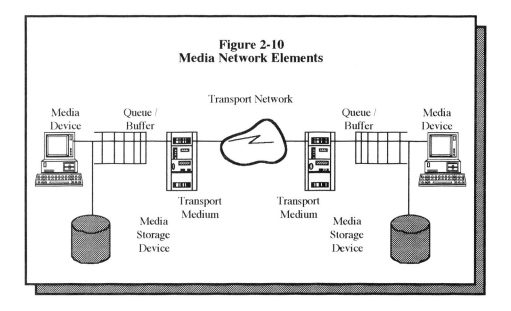

Figure 2-10
Media Network Elements

Summary

In this chapter, the various technology and engineering issues involved in the creation of multimedia systems is discussed. The focus of multimedia has to be the quality and the processing of the originating information, as well as its transfer to storage or other stations involved in the conversation. Therefore, a considerable amount of effort has been devoted to the resolution of compression, coding, and data transfer techniques. Recent attention has been paid to other issues such as data storage methodologies and transmission means.

References

(1)   Avedon, D., and J. R. Levy, *Electronic Imaging Systems*, McGraw-Hill, New York, 1994, pp. 20-50.

(2)   Gonzalez, and P. Wintz, *Digital Image Processing*, Addison-Wesley, Reading, MA, 1979, pp. 5-227.

(3)   Held, G., *Data Compression*, John Wiley, New York, 1983, pp. 18-57, 72-90.

(4)   McGrew, P.C., *On-Line Text Management: Hypertext and Other Techniques*, Intertext Publications, McGraw-Hill, New York, 1989, pp. 40-50.

(5)   Raymond, V., *An Introduction to Source Coding*, Prentice Hall, New York, 1993.

Chapter

3

Multimedia Technologies: Video Systems

Chapter Highlights:

This chapter is a continuation of the previous chapter that covered audio, imagery, and textual techniques for handling multimedia information. Video is an elaboration of still imagery and can be defined as motion-adjusted, time-compensated image sequences. Video processing techniques and technology have evolved in a haphazard way for several reasons, one of which is that, until recently, video has been the exclusive domain of the motion picture industry. As a result, cooperation between the computer technology industry and motion pictures has proceeded slowly. Advances in digital representation and compression of video sequences in recent years have made the computer control of video practical, and finally, high-capacity storage devices have made video handling and processing techniques available and financially practical. More recently, the issue of compression has become important, because of the limited bandwidth available to many transmission systems. Even coaxial and fiber systems raise concerns about bandwidth sufficiency due to the large number of channels that may be required for future market needs. The issue of video in multimedia is mainly concerned with the collection/processing, transmission, and display steps in the data usage sequence, as will be seen in this chapter.

3.1   Evolution and Importance of Video

The old adage that one picture is worth a thousand words is as valid today as when it was first uttered. Actually, nowadays each picture is worth even more, because a group of pictures, or a video clip, is usually accompanied by an associated audio segment, and the video is possibly linked to other language formats.   In addition, control and administrative information may also be embedded in each of the frames of the video clip, such as to what portion of the video clip the current frame belongs. One picture or a video frame can fill in many gaps and can answer many questions not possible with text narratives alone, while text narratives can be formatted as pictures as well.   Thus, practically all language forms are suitable for picture presentation.   Until relatively recently, still imagery, not to mention video, was well beyond the realm of the desktop workstation, or any other receiver medium, for many reasons, none the least of which included the technological constraints imposed upon the transmission system itself, i.e., the telephone system.

Technical Note 3-1
Historical Prelude to Video Teleconferencing

The original employment of twisted-pair copper wire for the use of telephone services was a decision based mostly upon its cost effectiveness rather than its potential for data handling, and eventual value as a wide bandwidth transmission capability. However, even during the early part of this century, experimentation into the application and utility of video was being conducted by such companies as AT&T's Bell Labs. In the 1920s, AT&T's Bell Labs, at that time conducting research at their facilities in downtown New York City, developed and tested the first 2-way, interactive video teleconferencing system which operated between 2 of its laboratory locations in Manhattan.

The current telephone system was never designed, let alone envisioned, to accommodate the requirements of 1-way, 2-way, or interactive video technologies, mainly because it was designed and implemented with limited bandwidth capabilities. The fact that video was either unknown or unrecognized at the time led to the negation of this medium as a future requirement. Thus, it appears that the current telephone companies have inherited a system that is sorely lacking in its ability to handle the needs of video information. They have been trying to overcome this miscalculation ever since. Attempts have been made to overcome these shortcomings with advancing technology, but there is a limit to such efforts and that limit is close at hand both in terms of technical capabilities as well as cost effectiveness. Unfortunately, interactive communication capabilities are exactly what telecommunications markets are demanding today, and are expected to demand in increasing bandwidths well into the twenty-first century. Technical Note 3-1 discusses the beginnings of teleconferencing as it was envisioned by early telephone system developers. Even then, there was considerable thought being given to the future and the integration of various media material.

An early attempt to develop a technology suitable for video transmission systems was researched by AT&T Labs. The history of early attempts is reviewed in Technical Note 3-2. More recently, digital picture and video imagery transmissions have become the major objectives for the era of fiber optics and satellite communications for several reasons. First, digital audio and imagery are less susceptible to corruption than analog versions. Second, digital information can be readily stored and retrieved in various storage media. Third, digital information can be more readily kept separated from other digital information streams than analog versions of the same information, thus maintaining the data's integrity and security.

Technical Note 3-2
Video Historical Perspective

Man has always wanted to visualize events that were occurring outside his/her immediate area of vision. First came the telescope, then binoculars, and later the microscope, still camera, movie camera, radar, video camera, camcorder, etc. Even early communications between literate, and even illiterate, people have been routed in imagery that has been recorded in some way and later relayed to the intended viewer. As technology progressed, the employment of binoculars brought man closer to the scene of interest, but only when telecommunications became a reality did our ability to interact at long distances become a distinct possibility. Video and associated imagery has always been a curiosity for humans, and has received varying treatments as to its possibility.

In movie serials of the 1930s and 1940s, video was depicted as having the ability to peer into all types of situations. Such movie idols as Gene Autry pursued armies from underground civilizations while his antagonists watched his every move through video monitors located in their subterranean headquarters. At the World's Fair in 1964, AT&T introduced the video phone. Its time was obviously not right, but the possibility has always been in the back of our minds ever since its conception. The video/telephone system idea has resurfaced in the past 10 years, stronger than ever, due to the availability of compression technology, wider bandwidth facilities of existing copper technology, and miniaturization of the various product elements. All of these events illustrate the public's interest in video over the years and its perceived importance to our quality of life. More important, movie and technology dreamers of the past have led the way into the future by pushing for the practical video applications that are present today.

Fiber optic systems of video transfer have the advantage of extremely wide bandwidth, although, fiber optic systems, like their copper cousins, have to be laid and strung physically throughout the area which they serve [2:1-11]. Fiber networks are beginning to encircle the major cities of North America as metropolitan area networks (MANs) and wide area networks (WANs) are implemented to carry broadband services. Fiber optics will be discussed later at length, but it also can support over 500 channels of TV and operates in the 1-micrometer (micron) range.

With regard to satellite systems, at least 2 vendors currently offer direct-to-the-home TV reception via satellite relay. The transmitter provides the multiplexing of many channels and power amplification for uplink to the appropriate satellite for rebroadcast to the ground receiving stations. The receiver stations employ an equipment array that consists of a small pizza-sized antenna dish, demodulation, decompression, subsystems, and a decryption unit. With this integrated capability the user can access up to 500 channels of video information and entertainment. At some point this system will be capable of interactive sessions where the user can access multimedia repositories.

Wireless technology has recently began to be seriously considered as a viable medium for CATV-like offerings. Such systems operate in the gigahertz (gHz) band ranges, such as X-band. Operations in these ranges allows the receiving antennas to be very small, on the order of 18 to 24 inches in diameter. These systems rely upon surface repeaters to guide the signal emissions around obstructions.

These wireless systems operate by having transmission antennas spread out in different locations [4]. The X-band range of these systems is very directional, with little bending of the electromagnetic wave. Thus, in high-density and high-rise building areas, receiving antennas must have repeaters available in many locations in order to provide the coverage needed for customers in different areas of the city. The bandwidth utilized by these systems carries the channels allocated for each system. Upstream signaling features for these systems have not been entirely worked out as yet, however, the 2 most likely options are to use the telephone system for narrowband signaling, and to provide an upstream capability using the basic bandwidth of the wireless system.

There are several problems that have to be worked out in order for these systems to be viable alternatives for CATV services, such as:

(1)   How reverse channel signaling, for interactive service offerings, is going to be provided reliably

(2)   How these systems can be operated in metropolitan areas

(3)   How multiplexing will be provided given the scarcity of the frequency spectrum

These wireless TV systems have the advantage of being ideal for high-density population areas, but suffer from the problems inherent in reflection and attenuation losses of the signals due to high-rise structures. In other words, antennas can be placed so as to receive X-band signals radiated within a downtown area, and several transmitters can be used so as to cover the required area. Sometimes, however, receivers on the fringes can be adversely affected and reflections can create ghosting and other distortions of the transmitted signals.

The recent major events in the arena of narrowband transmission technologies have been made possible by advances in the hardware that provides the stability and recovery of the generated signals, sophisticated coding algorithms, and the usage of application-specific integrated circuits (ASICS). Very-large-scale integrated (VLSI) chips, and very-high-scale integrated circuit (VHSIC) hardware developments are supporting ASIC progress, while the higher precision control and stability of generated wave shapes is made possible by very accurate versions of the phase modulation techniques employed.

Several video compression schemes have been developed over the past few years in the form of coding and transformation techniques. Initially, these techniques involved individual frames within the video sequence. More recently, frame-to-frame compression techniques have been developed such as those labeled as MPEG [9]. All of these will be discussed later.

At some point, neural network algorithms will no doubt be utilized to track objects within the scene as they move into and out of the view of the camera. This capability to track objects will also aid in the colorizing of black-and-white film, and will facilitate the compression of films by the use of vectors to direct these objects throughout the scene in progress. As mentioned in Chapter 1, several steps are involved in the coding of signals. These steps will be elaborated upon in this chapter.

All of the technologies discussed above point out the accelerating pace at which video is becoming an essential part of communications between humans. Now we will delve into the essentials of the technical requirements for a normal TV transmission.

3.2   Basic Video Parameters

Video as we know it today must be presented upon some sort of screen, either a TV or computer display monitor, or a flat plane array, in order to be seen and understood [1:6-1 to 6-15]. The flat panel display is coming into its own, due to advances in display elements and the processing units that control the displays.

The bandwidth capabilities of the video display device has to be sufficient to carry the signal. This is the first principal upon which video depends. Second, the size and resolution of the presentation should be as fine as that upon which the picture is based. A finer resolution in the display than that with which the picture was collected will not improve the picture resolution, but a coarser resolution in the display than the original collection will degrade the picture presentation.

There is a lot of confusion concerning video generation, presentation, and transmission issues. The basic problems with regard to the development and deployment of video have been the need to achieve a higher and higher resolution balanced with the need to provide transmission technologies that can achieve narrower and narrower bandwidths [10]. The network management of multimedia systems depends upon the digital replication of all the modalities involved. Thus, digital processing techniques for video are key to the handling of these important media. The topics involved in digital video processing and compression are discussed separately below.

3.2.1 Video Generation

Display imagery is created by a series of horizontal sweeps across a phosphorous-coated screen, during which time differing electron intensities, which are generated by an electron gun during these sweeps, create the image presented. The terms *display* and *video* will be used interchangeably in this discussion.   In North America, the North American Television Standards Committee (NTSC) standard is used. It specifies 525 lines per screen and is repeated 30 times, or frames, per second. The last 22 lines of each frame are not displayed, and this allows a range of capabilities to be included into the video display, such as:

(1)   The elimination of commercials from recorded signals

(2)   The incorporation of special audio and text information to be included into the signal stream

Each frame is composed of 2 fields, each of which consists of every other scan line within the frame and the other consisting of the other $1/2$ scan lines within the frame. The persistence of the phosphor coating on the inside of the display screen retains the first field until the second is traced so that the viewer perceives a smooth and complete image presentation.

The time required for each horizontal line is 63.5 microseconds (μs), of which 51.4 μs is devoted to actual information and the remainder (12.1 μs) is used for retrace of the scan line back to the starting position for the next horizontal line scan. The retrace time is referred to as the *horizontal blanking interval.* The *vertical blanking interval* is the time taken for retrace from the bottom of the screen back to the beginning of the screen trace at the upper left corner of the screen. This period requires 127 μs for the trip back to the top of the screen.

The 51.4 μs of the horizontal sweep is the period during which the actual display or video information is carried and impressed upon the display. This part of the signal is characterized by a series of zero-to-one transitions that represent black and white dots on the screen. These transitions replicate the familiar digital data stream. The density packing, or spacing between pixels, of digital information collected is dependent to some extent upon the bandwidth capability of the video monitor that displays the information received. The standard TV monitor has a bandwidth of less than 4.5 MHz, whereas, a computer monitor may have twice this bandwidth or more.

Many computer systems that display computer-generated text have been standardized on 80 columns with each column nominally consisting of 6 pixels in width and 8 pixels in height. However, modern word processing systems are fully capable of many other variations with regard to

(1)   The number of characters per line

(2)   The number of columns per page

(3)   Font types

(4)   Font sizes

(5)   Spacing between lines

(6)   Proportional spacing, etc.

The sequence of video images is composed of a series of frames of m by n rows and columns, respectively, and can be represented mathematically as $p_i[m,n]$, where **p** is a particular picture sequence nominally represented as a matrix, because it is portrayed as a 2-dimensional image on the display. The index i represents a particular frame within that sequence, while m and n are the pixel values for each row and column within each frame.

3.2.2 Video Presentation

Each horizontal scan has a video component that has a frequency, or rate of operation, and which is consistent with its pixel presentation requirements. Technical Note 3-3 provides a pictorial presentation of video generation. The reasoning for this is as follows. The video component of the horizontal signal has a pixel period referred to as T_p. This total period consists of the electron gun "on" period plus the "off," or wait, period that determines the spacing between pixels. The total number of pixels per line is N_p, and the relationship between these 2 parameters is:

$$T_p = \frac{51.4}{N_p} \ (\mu s \ / \ pixel)$$

where
$51.4 \quad = \qquad$ number of μs for the video data presentation portion of entire signal

$T_p \qquad = \qquad$ spacing or time between pixels plus pixel paint time (μs / pixel)

$N_p \qquad = \qquad$ number of pixels per scan line

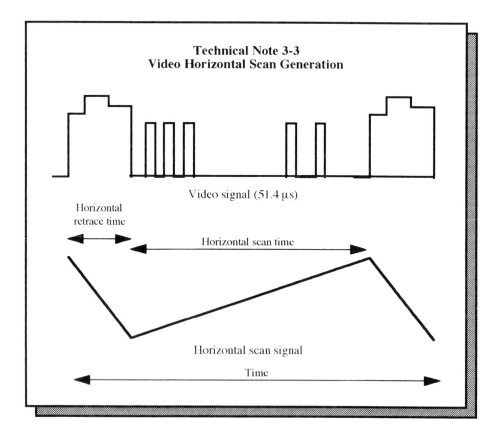

Technical Note 3-3
Video Horizontal Scan Generation

Video signal (51.4 μs)

Horizontal
retrace time

Horizontal scan time

Horizontal scan signal

Time

The video bandwidth is the most important parameter for video presentation, and it is calculated with use of the formula:

$$F_V = \frac{N_p}{100} \quad \text{(MHz)}$$

where

F_V	=	video bandwidth (MHz)
100	=	scaling factor used for conversion to MHz
N_p	=	number of pixels per scan line

Thus, if the number of pixels per scan line for a computer system is 480, the video bandwidth frequency has to be at least 4.8 MHz, i.e., 480/100. Since the video amplifier requirement of the standard NTSC TV set is considerably less than 4.5 MHz, an 80-column computer presentation cannot be shown on a normal TV screen unless that unit has a larger amplifier bandwidth. Older-style computer systems, such as the original Apple II, were restricted in the number of columns of characters that they could display, because the displays had insufficient bandwidth to support anything other than about 40-column displays. The vertical retrace interval, not shown in Technical Note 3-3, is that period of time needed to reposition the electron beam back to the upper left-hand portion of the screen prior to another frame or field scan.

Typically, the number of lines displayed is dependent upon the aspect ratio of the displayed system. The aspect ratio is the ratio of the length, or vertical height of the display, to the width of the display area. Thus, a ratio of 3:4 means that the length of the display is 75 percent of the width. In terms of pixels, this means that if the width is 480 pixels, the length will be restricted to 360 pixels, or a maximum of about 45 lines, depending upon the point size of the font used.

For graphics, on the other hand, not only is the number of pixels important, but also their density; i.e., the separation between pixels is key to an acceptable presentation. T_p, as explained above, is the pixel cycle time, that is, the time required to paint a pixel on the screen plus the interpixel off time. If the interpixel off time is decreased for a given scan time, the resolution of the display is improved, because more pixels can be displayed per line, and such graphic distortions as aliasing are diminished. The sharpness of the displayed image is significantly improved as the resolution, or spacing between pixels, is improved. The progress and continuing development of video has led to the possibilities of new video services similar in access to current telephone services. Two of the most important of these new services are discussed in the next section.

3.3   Emerging Video Services

3.3.1 Video Dial Tone

One of the key ingredients to the successful implementation of multimedia networks is the ability of the network to account for the usage of the resources utilized by the users of those resources [3:516]. First of all, there is the issue of how to allocate properly the bandwidth needed to transmit the various media involved. Second, there is the issue of how the sender can signal the receiver of his/her intention to initiate a multimedia transmission. This has been referred to as the *video dial tone* [7]. The term has become a political term used to herald the coming of a new technology era. In reality, video dial tone already exists in different forms, because once a communication link has been established, the circuit can be used for any media purpose, and in fact is used for different purposes. For example, a regular telephone call can be used to exchange data, video, and audio as well as a normal conversation. The emerging network management requirements are typified in Figure 3-1.

There are 2 principal ways that the various modalities may be utilized, one being the transmission of all the modes by separate circuits, and the other being the transmission of all modes by the same circuit where the broadband circuit supporting all the multimedia modes allows multiplexing of the media for joint transmission. Considerable work has already been conducted in the area of fractional digital services that are a subset of the T-1 standard of 1.544 Mb. These subsets include 256 Mb, 128 Mb, and 64 Mb, to name a few. The accounting management function of the network management architecture as set forth in the ISO standard for network management is the locale for this tracking activity, since the session setup will have to establish the conditions for the digital traffic interface. Indicated below are some of the parameters for this call setup.

Possible video call setup parameters are:

(1)   Dialed number
(2)   Bandwidth requested
(3)   Preferred carrier
(4)   Equipment vendor requested parameters, e.g., compression ratios, protocols to be used, etc.

The accounting function will trap at least the first 3 parameters, along with the duration of the call, so that proper cost allocation can be made to the appropriate customer account. The *station message detail record* (SMDR) system or the *automated message accounting* (AMA) system will have to be modified to collect all of these additional parameters.   Once the necessary software modifications have been made, and the needed hardware control has been set up to provide the fractional capabilities requested, the network will be ready to support the multimedia applications required.   Once both the fractional bandwidth and video dial tone have been established, multimedia can realize its full potential.

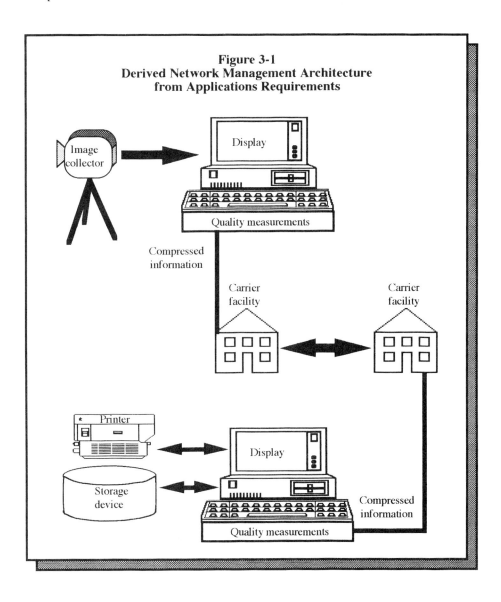

Figure 3-1
Derived Network Management Architecture
from Applications Requirements

3.3.2 Packet Video and Audio Accounting Systems

The other major ingredient of operational support is the utilization of digital technology to facilitate not only the technical transmission and storage but also the accounting of such transmissions. Asynchronous transfer mode (ATM), along with frame relay and other high-speed packet transmission technologies, were developed to provide the user with a means of sending and receiving packet audio and packet video [6:289]. These technologies operate in the ranges of 155 Mb/s to 622 Mb/s, sufficient bandwidth to accommodate these video needs.

Currently, the NTSC television standard transmitted in an uncompressed format requires about a 45-Mb/s bandwidth. As high-definition TV (HDTV) becomes more and more prevalent, its bandwidth requirements may require as much as 90 Mb/s to transmit. The bandwidth availability for ATM is more than sufficient to accommodate the requirements of either NTSC or even the HDTV standards. Since the ATM bandwidth is more than sufficient, it can be used for packet transmission of video services. A video packet is a subset of the entire video frame that has been subjected to analog-to-digital conversion, coded, and, in some cases, compressed before being packetized. Typically, each pixel is graded into an 8-, 16-, 24-, or even a 32-bit stream that will represent its gray-scale value. Most of the time, each pixel will be represented by an 8-bit byte.

The transmission of each pixel or a set of pixel values in each packet requires that each packet contain a header so that the packet can find its way to the proper destination. The header is used in 2 ways, as a means of addressing the packet to its destination, and to accumulate accounting information on the set of packets transmitted between 2 points. The network carrying the packets utilizes packet counts and addresses of these packets to account for the usage of the network facilities. Subsequent to this, the information may be used for billing based upon the distance and volume of the packet transfers. In this way, packet information is used for the accounting function.

3.4 Video Storage Technologies

Several technologies have been developed to accommodate the storage of video information. Some of these are discussed below [1:40-60].

3.4.1 Laser Disk Technology

The laser disk technology thrust enables the user to store extremely large amounts of data. There are still some significant cost issues involved in the creation of such storage media, and the equipment required is not readily available on a widespread basis. However, no doubt these problems will become nonissues as demand builds. This development activity has evolved into 2 main areas thus far, namely, write-once-read-many, or WORM storage, and compact disk-read only memory, or CD-ROM disks. Eventually, a write-many-read-many capability, similar to the conventional magnetic disk used in computers today, will take its place as the technology of choice for very high-density storage requirements.

The basic idea behind the optical storage systems is that as the disk is rotated under a laser, during the write process, the laser is activated and burn spots are created that represent data bits for later interpretation under the read heads during data retrieval. During the read process, the read head scans the spots on the disk surface by transmitting light onto the disk. The difference between ones and zeros is determined by whether or not the read head can receive a reflected signal.

3.4.2 WORM Devices

The write-once-read-many (WORM) devices allow the user to store permanently certain information, such as policy-holder history for an insurance company database, and to retrieve it as needed for future reference. This technology has been suggested for video content as well, where Hollywood movies would be permanently captured on this magnetic medium and accessed as required by subscribers as they request the movie library titles. The larger, 12-inch, WORM disks would have a capacity of 2 Gb each, enough storage for a typical movie. However, because of the fact that WORM systems have read-write heads on one side only, the platters would have to be turned over so that the storage capacity on both sides of the disk could continue to be accessed. This capability now exists. Additional engineering is required to assure that buffering is performed to maintain a constant video frame rate.

3.4.3 CD-ROM Technology

The CD-ROM technology consists of a player embedded within or attached to a suitable platform, such as a workstation, over an appropriate interface, such as SCSI. SCSI is a parallel attachment interface standard for devices that must be connected to computer systems. Such systems include storage devices, CD players, document scanners, etc. Even though CD-ROMs contain up to 600 MB of storage capacity, their retrieval speeds are slow by comparison with conventional magnetic disk capabilities.

3.4.4 Optical Disks

A more recent development in the realm of jukebox technology has been the optical disk. It is an outgrowth of the earlier audio disk technology development. Standards for its recording methods and internal disk organization have been evolving, but will no doubt be settled in a short time due to the commercial pressures being brought to bear. The costs for optical disk storage are less than magnetic systems, but the seek times are also much larger as well.

3.4.5 Magnetic Disk Storage Technology

Magnetic disk storage is a well-established technology that has the added advantage of being rewriteable. The basic approach here is to magnetize spots in one direction or another in order to establish the existence of ones or zeros at particular points along the surface track. Technical Note 3-4 illustrates the technique used to manage the reading and writing of magnetic disk systems.

3.4.6 Video Jukeboxes

The applications environment for video jukeboxes is to provide these units as servers from which programming can be distributed over physical transport

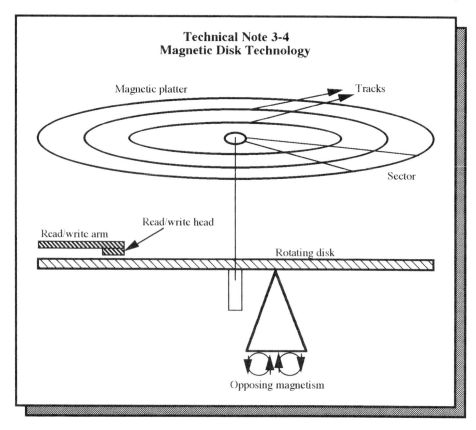

Technical Note 3-4
Magnetic Disk Technology

Magnetic platter

Tracks

Sector

Read/write head

Read/write arm

Rotating disk

Opposing magnetism

systems. The signaling, accounting, billing, and control over such applications is dependent upon the specific architecture and implementation techniques selected. For example, one type of video jukebox system might be developed for a housing development, while another might service individual homes widely dispersed over a large geographic area, such as the entire U.S.

The capacity of video jukeboxes is typically between 15 and 150 disks and about 5 disk drives. At the beginning of the operation of the disk drive, a robotics-style pick-and-place arm selects the disk slot specified by the control software, and that, in turn, was requested by the subscriber. The robot arm turns the disk so that the proper side is exposed to the read head, and slides it into the drive, also commanded by the control software. The access and manipulation times for this process may be as much as 15 s plus. Even though this is not very fast by computer engineering standards, the idea behind the video jukebox is to respond to human requests within a relatively short time period of indifference. This delay time is about what one would experience in initiating a VCR system.

Current efforts to compress movies for distribution over MPEG-based systems have not been very encouraging thus far insofar as the compression activities have been concerned. Even though the video jukeboxes have tera-bytes of storage capacity available, a typical movie requires over 2 Gb to store an uncompressed movie. The storage of 100 or more movies begins to tax the storage capacity of most any system, not to mention running up the costs involved.

3.5   Video and Image Collection

3.5.1 Video Cameras

Video cameras, or camcorders, can be used to collect and input video frames directly into the multimedia workstation. The standards supported in many such peripheral devices include PAL, SECAM, and NTSC. In addition, other methods are also supported such as composite video and S-video, both of which are of the NTSC variety. The difference is that the S-video is a higher-quality recording but when displayed on a monitor represents the same quality presentation as the normal NTSC signal.

The typical setup software for video input and recording has a variety of settings and configuration options available. Among these are the expected video settings that support color balance (hue), brightness, contrast, etc. The audio input and its digital format can also be adjusted in terms of the standard to be used, the number of bits per audio word, the choice of sources, and the dynamic range to be covered. If the input is S-video its input port may be a circular 5-7 pin plug, whereas if the input is a composite video, its input may be an RCA plug.

3.5.2 Document Scanning

Document scanning is a capability that has been around for some time, but only until recent cost advantages could be realized did document scanning, or electronic imaging as it is sometimes called, prove to be a viable possibility on a widespread basis [1:73-74]. This technology actually consists of 4 steps, these being:

(1)   The scanning of a document to convert it to electronic form so it can be manipulated

(2)   The segmentation of the electronic image into text and graphics

(3)   The compression of the segmented information

(4)   The storage of that data for later retrieval and reversal of the first 3 steps

Each step requires its own unique set of capabilities. The storage of scanned documents is similar to the storage of images, because scanned documents are treated the same as graphics and pictures. If the information represented in the scanned document consists at least partially of text, OCR can be invoked to convert the text portion of the scanned image to ASCII or other text format.

3.6   Video Processing

Like audio signals and still imagery, video compression relies upon a digital format and a sufficiently wideband transmission link to yield an acceptable result. A picture is a display, in some spatial coordinate system, of varying light intensities sometimes shown in colors of red, green, blue, or combinations thereof [5:5-220]. A video stream, on the other hand, is a set of such pictures closely associated in time, and characterized by a temporal redundancy in the imagery on an interframe basis. In other words, the objects seen within the video appear to move only very slightly between successive frames. Thus, there appears to be an opportunity for source coding in such a way as to take advantage of these slight movements. If the element of depth is added to each picture, the result may be a 3-dimensional presentation, as depicted in a virtual-reality presentation, a 3D movie, a hologram, or something else of the like.

Recent increases in the market interest and needs for video programming material have lead to increased compression, or source coding, development efforts in this area. The number of TV channels made available over coax to the home, for example, has increased from a nominal value of 36 to, according to predictions, over 200 in the near future, necessitating both new physical transmission means as well as compression techniques to assure sufficient bandwidth over the transmission medium.   Video teleconferencing, and its smaller brother, the video phone, will find increasing market acceptance soon, also requiring compression techniques to provide the functionality required. Finally, the wireless delivery of video and voice will require compression to accommodate the limited spectrum available, so that the increased load of channels can be supported. An example of this overall process is illustrated in Table 3-1.

Table 3-1
Calculations for Movie Sizes

Image Width by Height	640 x 480	640 x 480	640 x 480
Number of Color Bits (Image Depth)	2	8	8
Frames per Second	15	30	30
Compression Ratio	4:1	4:1	10.1
Compressed Image Size	2.304Mb	18.43Mb	7.372Mb

The storage and retrieval of video is a process that requires not only storage capacity but also sufficient speed of retrieval and decompression to be of value to the process so as to provide the imagery of interest to the viewer. Imagery storage requirements result from defining a combination of parameters, such as image size, image depth, i.e., gray-scale levels or colors involved, and compression ratio. These parameters are integrated into a function known as the *movie size*. The movie size is defined as the amount of space required for movie storage. It consists of the image size, spatial characteristics, etc., times the number of frames per second required for playback, times the number of seconds involved, times the compression ratio.

3.6.1 Fourier Transforms

One technique that could be used to reduce the requirements for bandwidth is that of reducing the image or each line of an image to a Fourier transform [2:45-59]. Each of the 525 lines of a normal NTSC video frame can be reduced to a Fourier transform. A Fourier transform is a representation of the signal of interest reduced to its constituent frequencies.
A 1- dimensional signal x[n], has the Fourier transform:

$$X(e^{j\theta}) = \sum_{n=-\infty}^{\infty} x[n]e^{-j\theta n}$$

Where

$X(e^{j\theta})$ = frequency component of the original signal, x[n],containing the frequency element θ

$\sum_{n=-\infty}^{\infty} x[n]e^{-j\theta n}$ = the summation of the constituent frequency

components of the original signal; $x[n]e^{-j\theta n}$ is the value of each frequency component

At the receiving end, the signal is reconstituted from its Fourier components. Each pixel is assigned its proper value based upon its position within the display line. Fourier transforms can be applied to 2-dimensional picture requirements as well, and a time-sequenced set of frames can be captured in a 3-dimensional Fourier transform. For a 2-dimensional picture, the transform would be represented as:

$$P(e^{j\theta h}, e^{j\theta v}) = \sum_{n=-\infty}^{\infty} P_i(n, m)\, e^{j\theta h}\, e^{j\theta v}$$

Again, the picture is expressed as horizontal and vertical components.

3.6.2 Neural Networks

If each video frame or each image frame could be coded in some way so as to capture the description and features of shapes within those frames, the bandwidth could be reduced significantly [3:4-6]. Much work has been done with neural

Technical Note 3-5
Neural Network Application to Network Management

In a network management situation, all, a majority, or the most important monitoring points may be presented to the neural network for analysis and resolution [2]. Such a simplified example is provided in Technical Note 3-6. Here a very modest network is replicated where two inputs are monitored, namely the signal-to-noise ratio between a microwave antenna link, and the block error rate for a packet assembler/dissembler (PAD) interface. A depiction of how this example can be monitored in the context of a neural network environment is shown in the figure below. Notice that there is again a range of values for both input parameters, F1 and F2, such that a conclusion about whether the problem is a source error, reception error, or an interface error falls within a range of parameter values. Pictorially, the results show up as areas within the domain of the problem. The shaded areas within each labeled area are the instances of the actual events. Those areas outside the shaded areas but within the labeled areas depict those instances of correct analyses that yield no valid results.

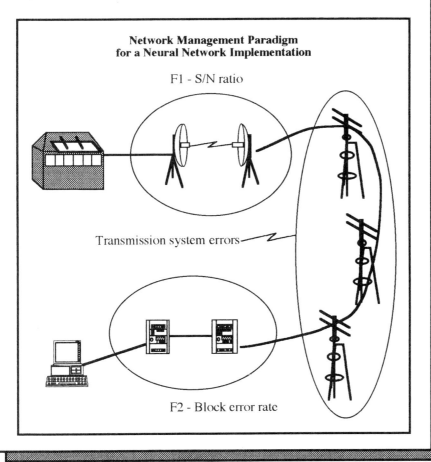

Network Management Paradigm
for a Neural Network Implementation

F1 - S/N ratio

Transmission system errors

F2 - Block error rate

networks within recent years to categorize and define various shapes or states of being for many different objects. Neural networks could define a feature space that tracks and replicates movements throughout a video presentation [8]. Technical Note 3-5 describes the background for a typical neural network problem where a limited number of parameters are to be examined. The explanation of how a neural network might be implemented is illustrated in Technical Note 3-6.

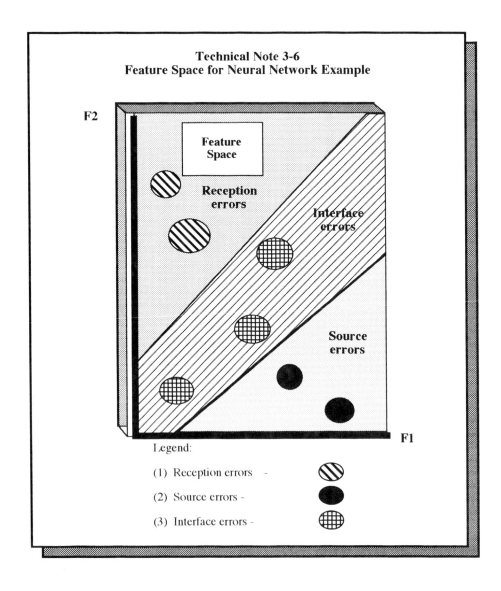

Technical Note 3-6
Feature Space for Neural Network Example

F2

Feature
Space

Reception
errors

Interface
errors

Source
errors

F1

Legend:

(1) Reception errors -

(2) Source errors -

(3) Interface errors -

3.7   Video Transmission Technologies

3.7.1 Fiber Optics

The basic technology architecture for a fiber optic system is for the incoming signal (transmitting end) to be encoded by some mechanism, such as pulse code modulation (PCM), after which the encoded digital signal is converted to light pulses [3:3-30]. This light stream is conveyed to its destination by a flexible glass tube referred to as fiber optics. At the receiving end, the light signal is detected and amplified. Following this, the light stream is reconverted back to electrical pulses and finally back to the original signal. There are several types of glass fibers that are used to convey the light signals including:

(1)   Multimode step index

(2)   Multimode graded index

(3)   Single mode

A full discusssion of the technical specifications of these fiber types is beyond the scope of this text, but the basics are as follows.
    The multimode fibers convey light sources to considerable distances without regeneration by refracting the light waves periodically as the signals traverse the length of the fiber conduit. Multimode fibers have a typical capacity of 10 to 300 Mb/s, with an overall bandwidth of about 25 to 30 MHz, over distances of 8 to 10 km (5 to 6 miles). Single-mode fibers have less of a refractive mode of operation and therefore can convey much higher bandwidths, but at a cost of shorter distances. Typically, single-mode fibers can carry at least 565 Mb/s signals to 5 km (3 miles) before being subjected to power regeneration.
    A figure-of-merit can be used to compare the relevant characteristics of fiber optic systems to twisted-pair copper wire and coaxial cable. The basic formula is:

$$\frac{\text{Cost}}{(\text{B x L})} \quad = \quad \frac{\$}{(\text{MHz x km})}$$

This figure-of-merit indicates the relative cost per bandwidth (B) combined with the transmission distance (L) for the particular medium. This ratio yields results, for the 3 principal transmission media, of about 15 for twisted copper wire, 18 for coaxial cable, and 1 for fiber optics. This ratio only indicates that fiber optics is the most cost-effective means of information conveyance because of its wide bandwidth capacity. Copper and coaxial cable has a much lower bandwidth and therefore a higher figure-of-merit. Even though fiber is about 20 times more expensive than copper, its tremendous data rate capacity makes it the medium of choice for video and large volumes of multiplexed data signals.

Multimode fiber is characterized by having comparatively large refractive angles as the pulsed light source traverses the glass medium. Single-mode fiber, on the other hand, has the feature of having a very slight refractive angle as the signal passes through the glass medium. The difference between these 2 fiber modes means that the multimode fiber has to be regenerated at shorter distances than the single mode fiber because considerable amounts of energy are absorbed each time the light source is refracted off the fiber wall. Figure 3-2 illustrates the differences between these 2 modes of fiber transmission.

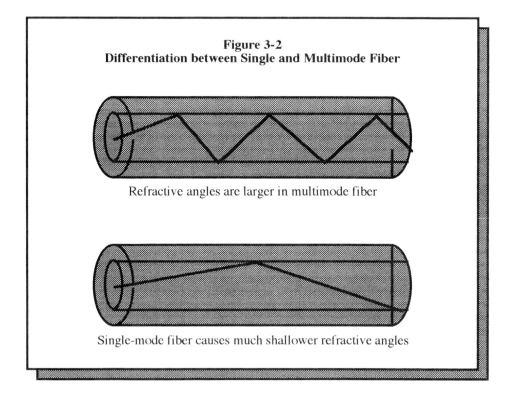

Figure 3-2
Differentiation between Single and Multimode Fiber

Refractive angles are larger in multimode fiber

Single-mode fiber causes much shallower refractive angles

Summary

The technologies used to create, process, transmit, and retrieve video information have been constantly evolving over the past 50 years. The advent of commercial TV has prompted the development of these and other supporting technologies that can assure the quality of imagery and/or video frames. It has also become increasingly clear since the advent of cable TV (CATV) systems that their abilities to assure consistent and ongoing quality of the delivered product left much to be desired. In a roundabout way, CATV has been responsible for the major thrusts in the areas of video delivery quality assurance.

Chapter Three

References

(1)   Avedon, D., and J. R. Levy, *Electronic Imaging Systems*, McGraw-Hill, New York, 1994, pp. 20-60.

(2)   Cannon, D. L., *Understanding Solid-State Electronics-Vol II*, Texas Instruments, Ft. Worth, 1985.

(3)   Edwards, T., *Fiber-Optic Systems Network Applications*, John Wiley, New York, 1989, pp. 3-30.

(4)   Goldman, J. E., *Applied Data Communications*, John Wiley, New York, 1995, p. 516.

(5)   Gonzalez, P. Wintz, *Digital Image Processing*, Addison-Wesley, Reading, MA, 1979, pp. 5-227.

(6)   Held, G., *Data Compression*, John Wiley, New York, 1983, pp. 18-57, 72-90.

(7)   Heldman, R. K., *Information Telecommunications Networks, Products, & Services*, McGraw-Hill, New York, 1994, p. 289.

(8)   Hoffman, M., "Technology Profile Neural Networks,"*Techmonitoring*, SRI International, July 1991.

(9)   Hsing, R., J. W. Lechleider, and D. L. Waring, "HDSL and ADSL: Giving New Life to Copper," in *Bellcore Exchange*, Livingston, NJ, March/April 1992.

(10)   Raymond, V., *An Introduction to Source Coding*, Prentice Hall, New York, 1993.

4

Multimedia Technologies: Multimedia Data Management

Chapter Highlights:
The requirements for multimedia data delivery systems differ somewhat from those of the more traditional data storage systems. Multimedia systems rely heavily upon the transmission of audio and video data elements from one point to another for use or appreciation by the recipient. Because of this, the information must not be discontinuous or sporadic in delivery, because the presentation at the receiving end may also be sporadic. Therefore, the multimedia data storage and retrieval systems have to be constructed so as to deliver the data at a constant or near-constant rate with sufficient memory to accommodate the entire clip of information. There are a couple of key issues that must be addressed and solved in order for the media storage system to be acceptable. First, the multimedia delivery system must be isochronous, meaning that the video or audio data must be transferred at a near-constant rate so that the presentation at the receiving end is also constant. Second, the system must have a buffering system so that the transmission system can deliver a constant rate of data elements. Third, the storage volume of the media database must be of sufficient capacity so as to accommodate all of the data involved, which includes data compression as well. This chapter will discuss these and related issues.

4.1   On-Line Storage Technologies

Magnetic disk technology is still the primary medium for the collection and storage of media information, although CD-ROM and laser disk technologies will likely replace magnetic storage for, at least, read-only operations in the foreseeable future. On-line, or peripheral, storage is primarily provided through the employment of magnetic disk technologies that utilize one or more circular platters to store the data of interest. These storage devices consist of one or more round magnetic surfaces that can rotate at very high rates of several thousand revolutions per minute (RPM), and can be stacked and ganged together for greater data handling capacity. Stacked platters provide greater storage capacity and shorter seek times than single platters. Storage efficiencies and storage speeds are enhanced by defining cylinders within platter groupings, where a specific track across a group of platters is defined as a cylinder. This cylinder can be quickly located by the set of read/write heads servicing the array of platters involved, and information written and/or retrieved can be more easily and quickly transferred.

Of equal importance to the facilitation of rapid storage and retrieval of information is the number of read/write heads available to the disk system. Typically, each platter will have its own read/write head, and in some cases, each platter will be serviced by multiple read/write heads. Multiple-head systems are important for multiple sessions of the same information, such as when there are multiple requests for the same movie. One subscriber may be 10 minutes into the movie, while another may be 27 minutes into the same feature. The number of simultaneous watchers will be somewhat limited by the number of heads available for information retrieval.

Personal multimedia creation and access to networked multimedia are dependent upon magnetic storage capabilities, because CD-ROMs cannot be used to write data. As a result, the effective utilization of magnetic media for multimedia purposes is key to the usage of this medium for multimedia purposes. This technology will be discussed in subsequent subsections.

4.1.1   Physical Organization

Each magnetic disk platter is divided into a number of *tracks* each running concentricly relative to the others that run around each platter at different radii from the center of the platter [4:10-40]. Similar *tracks* in a number of stacked platters define what is called a *cylinder*. A *cylinder* is an imaginary cylindrical tube passing through 2 or more platters at some specific track location. The *tracks* in each platter are, in turn, divided, by means of software partitioning, into *sectors*, and each *sector* is further divided into a number of the basic disk storage units, known as *blocks*.

Figure 4-1 portrays this partitioning arrangement.   A *block* consists, usually, of 256 bytes of data.   Each *block* has its own address so that the disk read/write heads can locate the *block* of interest.   The disk file management software locates a specific location by defining the *block, sector, track,* and platter, or *block, sector, cylinder,* and platter.   In summary, a group of *blocks* defines a *sector,* and there are some number of *sectors* within each *track.*   Each *track* follows a circular path about the center of the platter as a contiguous unit, and a series of *tracks* is located on each platter and is arranged in a concentric pattern.

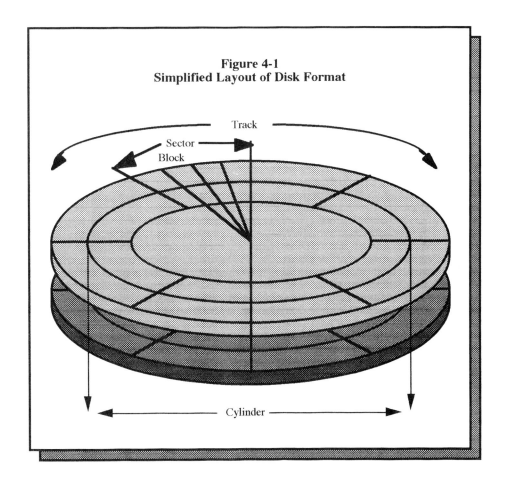

Figure 4-1
Simplified Layout of Disk Format

There are certain specific hardware components that make up the basic magnetic disk storage device. These hardware components that are responsible for the storage and retrieval of data from these devices are as follows:

(1)   Disk controller with host interface and buffer control

(2)   Read/write heads and associated amplifiers

(3)   Rotational speed and protection mechanism

(4)   The platters with their mechanical linkages and motor control

The read/write heads are positioned via hardware control signals that specify the exact physical address to the head positioning amplifiers from the disk controller. If data is to be written to the disk, incoming data sets are temporarily stored in an appropriate buffer until the write heads are positioned over the track to be used, at which point the data bytes begin to be interpreted into electrical signals and passed to the write heads for engagement with the surface as magnetic marks representing either a one or a zero.

Each physical data repository may contain one or more platters, or disks, and each contains a directory that stores, at a minimum, the names of each file, their starting disk location addresses, data sizes in *blocks*, last update, etc. Data *block* locations are numbered for identification purposes, and each location address is written at the start of each *block* location. Thus, each physical location contains the location address and its values at that location.

4.1.2   Traditional Data Storage and Retrieval Schemes

A data retrieval session requires that the data disk read head position itself over the proper disk area for information retrieval, either by some addressing scheme that can be translated into a physical address, or so that the location be identified by a physical address calculated directly within the software. Either method can be implemented either by direct or indirect addressing methods. Direct addressing means that the surrogate address requested by the external requesting algorithm is made known to the database system. The surrogate address is a *key field* or composite key that can then be used to scan the data file containing the record of interest. Indirect addressing means that the address of the location desired is found at some other location that can be made known to the requesting algorithm. The general concept of record identification is illustrated in Figure 4-2. In this figure the physical address, however obtained, is used to locate specific records. Additionally, one or more keys may be used to search for and identify records of interest. Once found, the attributes of the record are used for whatever purpose is intended.

Once this information is available, the disk is commanded by the requesting computer to seek the data set at that location. The read/write head(s) is (are) positioned to the appropriate location by reading each location address along each *track* and extracting the data at that location corresponding to the correct address. A whole technology has sprang up over the years to associate an address within a file or data set with a specific piece of data that may be needed.

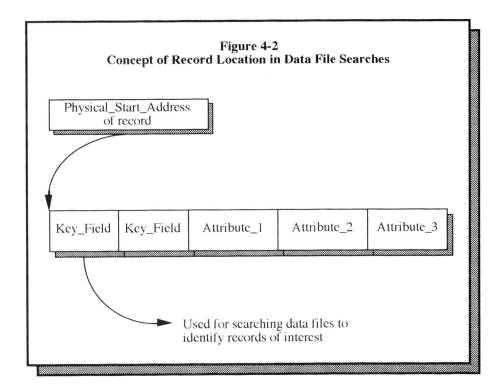

Figure 4-2
Concept of Record Location in Data File Searches

The storage of data is determined according to the file methods used and the software that provides supervision over the data involved. For example, the use of a *sequential file* storage method may, additionally, require some way of determining wanted locations within a file, without knowing these locations ahead of time. Technical Note 4-1 discusses the historical backdrop for data access methods. Several approaches have been devised to locate these required data locations, such as direct lookup, where the physical address of the data location can be specified directly, as opposed to indirectly, where a *key word* cross reference table provides a lookup to the actual location.

Magnetic disk storage can be implemented either by hardware or software means. Data, whether it be digitally encoded audio, text, graphics, or even video, can be stored according to the same formatting scheme, or its own formatting scheme as required for each individual application. Any scheme revolves around the notion of data *blocks* of, nominally, 256 bytes each. In some situations, data blocking may occur in larger chunks of data depending upon the medium being transported or stored. Each byte may represent a time-sliced segment of an audio wave form, a pixel value in an image space, or possibly an ASCII text character.

Technical Note 4-1
Types of Data Access

The sequential file method is an older method of file management that is organized and implemented within the program code of the application program, such that the defined fields, or subsets of the record, are organized within the software code. Every access operation requires that each and every record be retrieved until the record of interest is obtained. This means that, potentially, all records may have to be retrieved before the record of interest is finally obtained. Additionally, the records themselves are not delineated by a record call, but rather as part of a contiguous group of concatenated strings.

In more sophisticated file management schemes, such as the random access method, the fields are defined in a data dictionary outside the application program. This data dictionary permanently segregates the data sets so that calls to a record or data field may be made directly rather than sequentially, and all information within a record is made available simultaneously with such a call. Random access files may also contain key fields. This logical architecture can then reside upon various storage media, the most common of which is the magnetic disk storage medium.

One of the standard and popular measures of the efficiency and compactness of programs and data files is their storage requirements. Any storage medium has a nominal storage capacity based upon the space required to imprint a single bit onto that storage medium. This nominal, single-bit storage space is multiplied by the total run length of the available storage medium, yielding the value of the nominal storage area. This value is advertised by vendors of these media, for instance, makers of $3^1/_2$-inch disks advertise their products as having either a 775-KB or 1.44-MB capacity. However, after they are formatted, these storage devices will only store a fraction of this capacity because of the formatting restrictions in the total storage capability.

Once these devices are formatted, their storage capacity is determined by a series of constraints that will be described as follows:

(1)   The distance between succeeding stored *blocks* is referred to as the *interblock gap*. This gap allows the disk program to distinguish one *block* from another based upon the physical distance between these data groupings.

(2)   Each *block* of stored data has its own unique address that accompanies each data *block*. The contents of each *block* then consists of the address of that *block* followed by its value.

(3)   Within each data *block* there is an *intrablock gap*. The *intrablock* lies between the *block* address and the *block* value.

(4)   The *key count* is the space made available within the block for the key. The key is a data reference that is used for searches made by the database software.

(5)   The *data count* is the value of the *data block*.

Once the storage scheme is worked out, the next major issue to be tackled is the query method to be used. The query methodology for multimedia is a request for the information clip to be examined. Unlike the traditional database query, which involves several questions and answers, the multimedia query is likely to be a single request. The nature of query-response interactions is covered in Technical Note 4-2.

Technical Note 4-2
Database Query-Response Approaches

The traditional query-response facilities of current database management systems provide the user with the ability to specify and retrieve references to key word(s), or requests, specified by the user via the query facility of the software. The response is usually a set of instances or references to the key word, or request, that are listed in some sort of predefined order. Query-response facilities have little or no intelligence built into their processing, and as a result, each piece of information retrieved has little relationship with adjacent information clips except for the key word association. Hypertext and hypermedia go beyond the query-response paradigm to the point of assembling the information clips into a coherent discussion of the subject matter.

Another very important issue associated with multimedia information systems is that of the storage capacity for a typical disk system. This issue is explained as follows:

$$C = \frac{S}{(i + b + n + k + n + d)}$$

where

C	=	formatted storage capacity
S	=	nominal disk capacity
i	=	interblock gap
b	=	block address count
n	=	intrablock gap
k	=	key count
d	=	data count

Of equal importance, especially when transferring images or video frames, is the information transfer rate. This value is determined by several factors, these being the seek time for the correct track location within the disk space, the positioning time for the read/write head, and the read/write speed of the platter. All of these parameters are related to the data transfer rate as follows:

$$ETR = \frac{B}{(R/2 + P + B/s)}$$

where

ETR	=	effective transfer rate (blocks per second)
B	=	data block size (bytes)
R	=	disk rotation time (milliseconds)
P	=	read/write head positioning time (milliseconds)
s	=	read/write transfer rate (bytes per second)

A typical video frame might consist of 640 pixels across by 480 pixels high, for a total image size of 307.2 KB. If a video frame contains 307,200 bytes of image data, then, assuming 30 frames per second for a video transfer, the nominal transfer rate will have to be at least 9,216,000 bytes per second. **B**, the data block size, consists of a certain number of pixel data values per block. For instance, if each pixel represents 1 byte of data, then a block of 480 bytes represents 480 pixels. If each pixel encodes 2 bytes of gray-scale data, then only 240 pixels are represented.

The data block size has to be large, as does the read/write transfer rate in order for the ETR to approach the rate of the nominal read/write transfer rate, **s**. For contiguous areas of storage where repositioning times for the head and seek times approach zero as limit, the read/write transfer rate is sufficient to accommodate the requirements for video transfer, since s is designed to match the actual rate required for video support. Compression and other encoding techniques may reduce this requirement, however.

4.1.3   Video Information File Structuring

The typical video file is referred to as a binary file, because it contains a stream of ones and zeros, not arranged in any coded format that corresponds to a text coding system, such as ASCII. The data stream may be compressed and therefore arranged according to some coding scheme, but it does not follow some type of text format. The problem that has to be addressed when dealing with how to store and retrieve such a file is how to segment its data within that file structure. The video message or clip may reside in one or more files, each of which has to be segmented for data transfer and decoding purposes. The segmentation process may occur at any of several points, such as by frame, by field, by line, or by pixel. If, for example, each frame in a video clip were to be transferred from storage to temporary memory buffer storage while being decompressed and decoded prior to relay to a display buffer, that frame may require considerably more storage than if video data were processed line by line. Figure 4-3 shows a typical binary file structure arranged on a line-by-line basis.

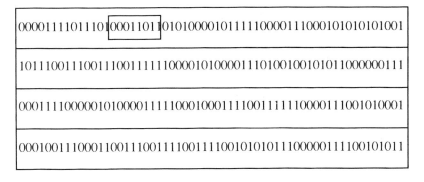

**Figure 4-3
Video Binary File Segmentation**

000011111011110100011011010100001011111000011100010101010101001

101110011100111001111110000101000011101001001010110000000111

0001111000001010000111110001000111100111111000011100101010001

0001001110001100111001111100111110010101011100000111100101011

Note:
The binary sequence framed in the first line may be interpreted as an escape character if processed as a text file and therefore generate undesired consequences.

As noted in the figure, if a video file were to be interpreted as a typical text file, it would eventually come across a sequence that would cause some type of unwanted response, such as an escape sequence. Therefore, the file is treated as a special case where each grouping of bits is assembled and interpreted as a binary sequence only before and after decompression and decode processing.

4.2   Attributes of Multimedia Database Systems

Early on in the development of multimedia facilities, such as those marketed by Apple, imagery and video were given their own file structures with the necessary designation and formatting capabilities, and later with compression capabilities. Thus, the search for and retrieval of a video or still frame required only the identification of the file in question.

Database systems that are available on the market today are a combination of proprietary features as well as those dictated by standards. As such, each product is sold on the basis of its own peculiar attributes, a complete discussion of which is lengthy and cannot be covered here [4:40-55]. However, the basic attributes of any of these can still be reduced to some elementary parameters that are discussed below.

4.2.1   Speed

The speed of a multimedia database system is a subject associated with both its input/output technology, i.e., its speed of physical data insertion and extraction or reading and writing, and its functional technology, i.e., the operational speed of the database software algorithms [4:46].   A database system is usually procured on the basis of its operational speed as well as its features, which includes initiation, manipulation, and update of the database system.   A multimedia data storage system is also dependent upon its speed characteristics, and on its ability to address a large number of different formats as well.

The output speed at which a data item may be extracted from its storage medium is important, not only for the immediate information needs but also for the overall performance of the system. If the system is not capable of rapid access and storage speeds, the overall system speed will suffer and limit the overall system performance.   Since the storage media of today are predominantly some type of disk, their attributes will be addressed here. The typical database system today is organized under software control in the majority of cases, thus leaving the exact details of addressing, formatting, and data extraction up to the individual vendor. The database system basically must carry the data address, the data or record itself, and one or more keys that are used to enhance searching. The time to retrieve one block or record data item can be represented by the following equation:

Access time (AT) = read/write head positioning time +
average rotational delay +
data read/write time

$$AT = \frac{R}{2} + P + \frac{B}{S}$$

where:

AT	=	data access time
R	=	disk or platter rotation time in milliseconds (division by 2 allows for the fact that on the average, it takes half of that time to reach the record or block of interest)
P	=	average head positioning time in milliseconds (also called seek time)
B	=	data storage block or record size (bytes)
S	=	nominal read/write time (bytes/ms)

The rate of transmission can be derived by dividing the access time into the block or record size, B. Thus,

$$\text{Transmission rate} = \frac{B}{\left(\frac{R}{2} + P + \frac{B}{S}\right)}$$

The speed of storage and retrieval includes read/write time, as previously discussed, plus algorithmic calculations for determining the read/write addresses, and the data entry time. All of these parameters contribute to the overall speed with which the database system executes its jobs. The data entry time is included in the man-machine interface issue as an additional factor in the operational speed of the overall system. Operational speeds are affected when the user is delayed in retrieving data items of interest through such problems as ambiguous display messages, confusing display graphics or tables, limiting features in the database system, and the like. These types of delays can cause an increase in the training time, capability required to perform the job, and, possibly, number of personnel needed. In other words, the cost of performance may be more a problem of the human interface than of the hardware or software.

The algorithmic calculations required to derive an address needed to write to a location or to read from one are a function of the sort and search techniques used in the implementation of the software system [4:70-90]. Following is a representative list of sort techniques and a rough comparison of their execution speeds. These "on-the-order-of" values O(expression) provide some indication of the relative speed with which these algorithms can execute some type of sort on N items.

Sort Type	**On-the-Order-of Values**
Binary	$O(Log_2N)$
Selection	$O(N^2)$
Exchange Selection	$O(N^2)$
Insertion	$O(N^2)$
Bubble	$O(N^2)$
Quicksort	$O(N\ Log_2N)$
Heapsort	$O(N\ Log_2N)$
Tournament	$O(N\ Log_2N)$

4.2.2   Capacity

Capacity is a function of the physical aspects of the multimedia storage medium. The storage capacity of a system is not as important today as it was just a few years ago [2]. The reason for this is 2-fold. First, hardware data bus technology has made it possible for several disk storage systems to be locally attached to the same computer simultaneously using a data bus architecture known as SCSI. This feature allows a primary plus a large number of auxiliary storage devices to be accessible to the computer at once. The SCSI architecture provides a design such that each auxiliary unit can be daisy-chained from the computer to each of the next devices in turn so that a string of hardware facilities is available for use by the system. Second, compression has made the utilization of storage space on a disk as cost-effective as possible. Compression algorithms can account for between a 1:4 and a 1:10 or better compression of data.

The nominal or theoretical maximum capacity of each track on a disk storage platter is reduced by the overhead usage that is necessary to separate and distinguish data storage from administrative and control requirements. Data are stored in conjunction with several categories of overhead information. The smallest unit that can be independently accessed is the data storage block. Within the block are the data subblock, which contains the actual information of interest, the address subblock, which identifies the marker for retrieval of the block containing the data of interest, the key subblocks, discussed previously, which provide indicators for specific search fields and index table entries, and possibly other administrative control subblocks.

Each stored piece of data is organized such that address subblocks, data subblocks, any key subblocks as well as the appropriate gaps in between are written on the disk as required by a particular program. The amount of actual data, not including wastage, extra bytes for gaps, and administrative data, is going to be something less than the theoretical maximum available. This available space is calculated as follows:

$$n = \frac{S}{(A + B + k + G)}$$

where

n	=	blocks or records that can be stored on a track
S	=	number of bytes nominally available for storage
A	=	address subblock
B	=	data subblock
k	=	key(s) subblock(s)
G	=	gaps associated with each subblock

4.2.3   Multimedia Information Buffering

Capacity also has another meaning when we talk about input and output buffers. The capacity of buffers is important to the flow control in a communications system so that certain points do not become overloaded. The design constraints for buffer capacity are usually expressed in percentiles. This means that the buffer is being designed so that some percentage of the time, the buffer will not overflow. This value is set normally as high as possible so that very high assurance of overflow avoidance can be expected. Values for the percent of the time that a buffer will not overflow range from 90 percent to 99.5 percent as a rule. The method for these calculations is as follows.

For a single multimedia service facility receiving video frames, a buffer capability is necessary in order to reconstruct incoming information prior to actual display. As will be seen later, a firm delay period is necessary in order to assure statistically that all pixel packets are received before the display is complete and ready to be output to the monitor. As the delay period increases, more buffer space and lag time is needed to provide a continuous video presentation. As the variance for the delay increases, more buffer space is also needed. The size of the buffer space for a certain number of video frames is:

$$\pi_{Bq}(95) = Bq + 2\sigma Bq$$

where

π_{Bq} (95)   =   95th percentile confidence interval for the
                                                image buffer size required (95% confidence
                                                interval for maximum queue length)
B_q =   average number of image frame buffers
                                                required (average queue length)
σ_{Bq} =   variance of the number of frame buffers
                                                required (variance of queue length)

Thus, a buffer requirement for a group of frames is determined by the length of the average queue and the variance of that queue length. If the average number of buffers required is 3 and the variance of that requirement is 2, then the 95% confidence interval for the actual number of buffers required is 7.

There are a number of issues related to the retrieval of what is referred to as continuous media requirements. These are summarized briefly as follows.

Playback -   Video retrieval and playback consists of a series of tasks that are executed to retrieve information bundles from the storage medium and strobe them out to the intermediate buffer storage unit for eventual relay to the presentation unit.

Storage -   The storage medium will be tasked with the capability to store hundreds of movies. Each of these will be stored in some sort of compressed format compatible with the application program.

Transmission -   The conveyance of the data stream from point to point requires a compressed format which will be reversed upon receipt at the receiving end.

4.2.4   Ease of Use

Ease of use is equivalent to the human interface necessary to make use of the system features [5]. Here it also includes everything associated with file management initialization, maintenance, and operations. In addition to data insertion and retrieval, ease of use is the other major ingredient that determines the speed of the systems. Complex skills may sometimes be needed by the human operators when performing updates, inserting data into the system, generating reports, or creating data dictionaries. These necessary skill sets may be traded off against better human interfaces that in turn may negate certain skills when considering the costs and utility or ease of use for the database system under consideration.

The number and quality of the workforce necessary to maintain and operate the system, may, in turn, drive operating expenses up significantly. Again, however, such costs have to be traded off against capital investment necessary to acquire such facilities. Ease of use includes such simple qualitative issues as the ability to use the screen presentation effectively, to make use of the data presented, and to jump from one functional screen to another as needed with minimum deliberation and effort. Quantifying the ease of use is difficult since it is an elusive parameter, but to some extent ease of use can be thought of as being very similar to group communications, discussed in Chapter 6. The more involved the interface is, the more that interface is like interacting with a large group. The message here, also, is that less time will be spent performing other productive tasks if more time is spent working with the operational software.

4.2.5   Processing Environments

One of the ways of distinguishing processing environments is to categorize them as being either centralized or distributed [4]. Processing activities do not of themselves dictate the database requirements of the system, but the operational activities as reflected in their architectures do provide some very good indications of how the databases are going to be used. If, for example, the requirements indicate that the database will not be accessed frequently during a normal operational period and the processing activity time is not critical in nature, then we may have a centralized environment whose operations call for a central repository of information that can be used and updated in a non-time-critical fashion. The database system to be used may provide housekeeping facilities, such as overflow and updating, that are not as sophisticated and fast as would be required if the requirements were more time critical.

If, on the other hand, pockets of processing are scattered around the system and local databases are efficient, then we may have a requirement for a distributed system that might consist of islands of processing linked to each other by some form of networking fabric. There are, of course other types of systems, such as hybrid systems, that consist of processing environments scattered around the system, but which also have centralized processors to act as database repositories for these islands of processing. As always, a detailed assessment of the requirements is a prerequisite for describing the data system design. The quality of the ultimate product is dependent largely upon the correct definition of these user requirements, as they are tailored to meet the user expectations.

4.3   Features of a Multimedia Data Management System

In any kind of data management implementation, there is a sequence of steps that must be taken in order to achieve operational stability [4]. These steps are outlined below.

(1)   Data Dictionary -

> A data dictionary must be established both in readable form as well as in code in order to organize and track information that flows into and around the system operation. The data dictionary describes each record type along with its various fields or record segments. The dictionary lists the fields in each record type, the definition of each field, its format, the authority for modifying the fields, and how or where the records are to be used. The software system supporting the data system must be capable of initiating and formatting data according to the format requirements described by the data dictionary. The data dictionary, upon being told that a certain piece of data is associated with a certain data type, will automatically format input data in keeping with the format requirements of that data type.

(2)   Applications Shell -

> A shell has to be implemented and utilized that organizes and assists in the retrieval of data while it is being inserted, reorganized, or deleted. There are several approaches to the implementation of such a capability. These include such approaches as organizing the data in a binary tree format, linked list, or some other arrangement so that there are facilities for handling data insertion, overflows, retrievals, and deletions.

(3)   **Human Interface -**

> The human interface must provide the facility for the user to insert and make data retrievals based upon input requests. It is display and input oriented such that the operator can make requests of the system easily and follow those requests as they are acted upon. Part of this interface is the facility for generating reports that may be requested as part of the query process. Formatting the output or being able to represent data graphically is also provided in this interface in most cases.

(4)   **Language Interface -**

> In most cases, also, a language interface should be provided so that the data system can accept specialized requests for data organization, retrieval, or output. This facility is especially important if the data system is to link with some other data systems or specialized hardware.

These are the basic elements of any data-worthy management system. The criteria for selection of such systems include access times, processing speed, data storage capacities, cost, and systems support. While these are important features, other very important considerations are the data dictionary interface, report generator, query facilities, software interface with other programs, restructuring abilities, error handling, documentation, support, and so on. Specific needs in particular situations may revolve around the questions of what equipment is already in place and what capabilities are easiest to acquire or use. The essential features cited above should, however, always be present.

4.3.1   Database Systems Interfaces

Database interfaces are of two principal varieties, either human or program derived. Human interfaces to data systems determine their compatibility and ultimate utility, because if the user finds the data system difficult to manipulate, it may be ignored or abused and possibly have little value in the network management process. Additionally, if the conversion of command languages or algorithmic interfaces from other software programs is inadequate, the data system may not respond to operational needs or may respond erroneously from processing bugs. Interface requirements should be detailed to the extent possible, with no surprises involved. The database system specifications that follow these requirements dictate the capabilities of the data system, for the user and supplier alike to see. These specifications are the result of considerable work to match industry capabilities to user needs and are critical factors in the go/no-go decision for system implementation or upgrade as well as the system's utility.

4.3.2   Complexity and Compatibility

Database systems that are targeted for time-critical operations such as network management tend to be complex in implementation but must also account for anticipated additions to the system in future upgrades. There is no simple solution to the problems implied by these concerns. The users and owners of these systems have to be alert to the consequences of failing to consider such problems as complexity and compatibility. The best way to attack complexity is to plan as far ahead as possible, to reduce every element to its lowest level of detail where practical, and to constantly look for ways that complexity can be reduced to straightforward actions. Compatibility could be expressed in terms of tangible features, such as physical, architectural, functional, and operational descriptions. *Interoperability* is one of the more popular terms associated with these issues. No matter what name is used, any form of compatibility starts with a properly defined set of database requirements.

4.4   Example of an Image Database in Operation

One of the first and large scale examples of image database usage, and possibly the rudimentary first step in the capture of data for multimedia databases in the multimedia experience for the user, is the Yellow Pages phone book. Recently, laboratory demonstrations of electronic Yellow Pages systems have shown the potential for database service providers to deliver Yellow Pages via Internet or other services, such as Prodigy, Compuserve, America On-Line, etc. Electronic Yellow Pages systems deliver text, and graphics as well as video clips to the viewer. The theory is that video adds more impact to the presentation.

The divestiture of the local exchange telephone companies from AT&T in the U.S. in the mid-1980s led to the ownership of the affected phone books being passed to the local exchange companies. Aside from the fact that the Yellow Pages business is a very high-margin money maker, the negotiations during the divestiture favored local control using the logic that local customers needed a local vendor. The production of the Yellow Pages evolved over the years to the point that now its features are part of the best publishing software packages available. The major difference is that the image database must be very large and accessible from many stations. The makeup of the Yellow Pages production system is substantially as follows.

(1)   Image Database -
The advertising artwork is created or updated either from existing electronic images or from hard copy generated by an artist. In the former case the graphics artist calls an existing advertising image from the database. This call is based upon a tag, or classifier, assigned to the image when it was originally captured. The operator's system is fully capable of graphics displays and is equipped with software to allow the operator to modify, crop, paste, etc., the artwork, and to restore it back into the image database. The database system has the requisite directory, and image compression and storage algorithms to allow the imagery to be properly maintained. In the case of preexisting hard copy, a scanner is used initially to capture the artwork. During this process, the operator is asked to classify, or identify, the scanned object. There are usually several image storage formats that the operator must choose among.

(2)   Type Setting -
The page layout is accomplished using special-purpose software to support pasting, cropping, and scaling of the images destined for each page. Initially, the software will sort through the images and/or advertisements to be used to perform an initial automated arrangement of the ads involved. Following this first-pass automated layout, each page can be reviewed in some sort of page preview format for quality control checking. Technical problems as well as customer wishes can be reconciled at this point, at which point the pages can be printed using a high-quality printer and sent to the printing company for copying and binding.

(3)   Video Inclusions -
The electronic versions of various Yellow Pages offerings can be equipped with windows at different places within the advertisement where video clips can be played while the user views the page.

Summary

The technologies for data management are wrapped up in the requirement that video and audio information must operate in a continuous mode. This boils down to the fact that the video delivery system must transport the information in a nonblocking environment and that sufficient buffer space must be provided to assure that the continous flow of video and audio will be delivered. This chapter discussed the various technologies associated with the storage and delivery of the multimedia information.

References

(1)   Gonzalez, P. Wintz, *Digital Image Processing*, Addison-Wesley, Reading, MA, 1979, pp. 5-227.

(2)   Held, G., *Data Compression*, John Wiley, New York, 1983, pp. 18-57, 72-90.

(3)   Hsing, R., J. W. Lechleider, and D. L. Waring, "HDSL and ADSL: Giving New Life to Copper," in *Bellcore Exchange*, Livingston, NJ, March/April 1992.

(4)   Loomis, M. E. S., *Data Management and File Structures*, Prentice Hall, Englewood Cliffs, NJ, 1989, pp. 98-180.

(5)   McGrew, P.C., *On-Line Text Management: Hypertext and Other Techniques*, Intertext Publications, McGraw-Hill, New York, 1989, pp. 40-50.

(6)   Raymond, V., *An Introduction to Source Coding*, Prentice Hall, New York, 1993.

Chapter

5

Multimedia Technologies: Transmission Systems

Chapter Highlights:

The transmission systems required for multimedia applications are of 2 major types, one being those that have the bandwidth necessary to support the basic service requirements, and the other being narrowband services that are grouped together for the purpose of achieving the bandwidth necessary. An additional technology that fits the use of narrowband services is information compression. Compression is used to achieve bandwidth capabilities that would otherwise not allow the use of the application in question. In the case of true broadband technologies, the one that will support future multimedia applications the best is ATM, which is implemented over fiber optic transmission lines. Of the narrowband choices available, the use of the inverse mux (to be explained later) will effectively multiplex narrowband links to yield broadband services. Compression also expands the effective bandwidth of the transmission link. Compression ratios from 20:1 to 40:1 are not uncommon. Analog technologies have also been proposed and in some cases used to deliver true interactive multimedia services. These are also discussed in this chapter.

5.1   Transmission Systems for Multimedia

Multimedia must take advantage of all sorts of transmission systems, both physical and functional. The physical systems consist of electronics plus the transmission medium itself to achieve the desired capability. The functional systems consist mainly of a packet protocol coupled with a fiber transmission medium. Some of the more interesting or important transmission schemes are outlined briefly below.

5.1.1   ADSL/HDSL

Recent developments in coding and modulation techniques have given new life to twisted-pair copper wires. Some of these innovations are known as ADSL or HDSL. ADSL is a technology aimed at providing compressed video at 1.544 Mb/s in one direction plus a signaling channel in the opposite direction. It was developed to support the telephone companies in their efforts to create transport facilities for the delivery of movies to the home as part of their positioning to compete with the cable TV (CATV) companies. HDSL was developed to provide simultaneous 2-way video over the same twisted-pair system [4:4-6].

5.1.2   ATM/SONET

Several packet switching technologies have emerged in the recent past that provide transmission and data control over the first 3 layers of the OSI model. One of these is known as frame relay, but it has been leapfrogged, in terms of utility potential, by the second major type of high-capacity packet technology, known as asynchronous transfer mode (ATM). It was originated by the CCITT as that body was developing its broadband integrated services digital network (B-ISDN) strategy [1:208-233]. There is an ATM forum that actively supports the development and manufacturing of ATM products. ATM/SONET is seen as the next-generation architecture for both public networks as well as private WANs. The vision is for ATM to be used as a building block where ATM LANs can be linked through WANs using telephone lines.

ATM is a virtual-circuit, packet-switched, networking system technology that takes information packets that have already been parsed and packetized at their originating points, and then repackages them by the ATM protocol to include a header for each packet that designates the routing information needed for each packet. The ATM paradigm is to take 48-byte groups of incoming data streams and to attach a 5-byte-header to each 48-byte group for routing purposes. SONET is the transmission and framing aspect of the ATM/SONET system. It provides an OC-3 (fiber optic transmission standard), which is a 155.55-Mb/s light modulated signal. A follow-on to OC-3 will eventually allow evolution of data rates up to 600 Mb/s or higher [2:380-430].

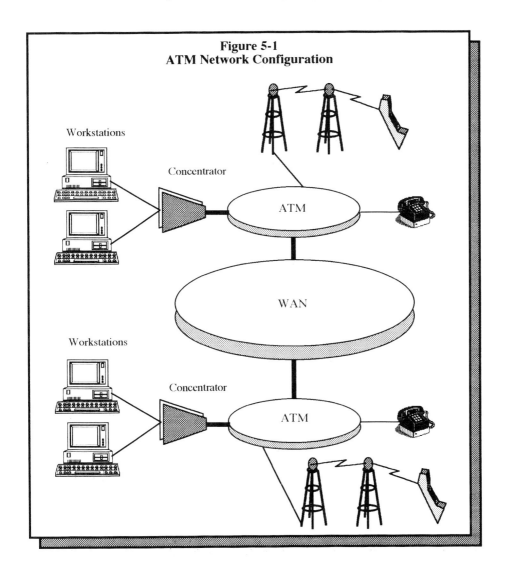

Figure 5-1
ATM Network Configuration

The fixed-length packets of ATM guarantee a fixed-length response time for applications, and therefore, it is an all-media support capability. Additionally, fixed-length packets help to assure minimum delay through the system. The nature of the packets may be either repackaged packet protocols, such as X.25, or raw data, such as video pixel information. In the former case, errors can be recovered depending upon the ability of the encased protocols to allow retransmissions of missing packets. Raw data transfers where pixel information is being transported, with only ATM header information available, may experience a dropout occasionally as packets are received in error.

ATM, therefore, is intended to allow users to access the ATM facilities at any point, and to have a guaranteed bandwidth with not-to-exceed delay limits during the time of usage. The initial offerings of ATM will allow users to access the network via multiplexers that will provide the gateways for ATM usage. A simplified portrayal of this approach is provided in Figure 5-1. In some cases the users are operating workstations and exchanging data files, using desktop video facilities, or exercising conferencing software to exchange ideas and information. Other potential users are accessing the ATM LAN using the telephone network as an interface, or gaining entry into the ATM network via some sort of wireless capability, such as cellular, trunking radio, etc. The session, or conversation between these parties and other users elsewhere in the system is provided by access through a wide area network, facilitated by fiber, and further interfaces to another ATM LAN for termination at the desired location.

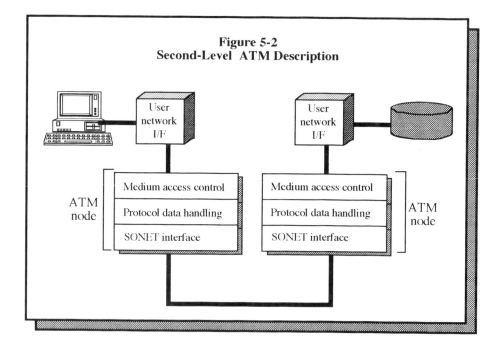

Figure 5-2
Second-Level ATM Description

The details of the ATM architecture are now going to be elaborated upon. Figure 5-2 shows this second level of detail. At the customer premises equipment (CPE) point, the user must have an interface to the ATM network that is referred to as the user network interface (UNI). This combination of hardware and software determines the nature of the transmission, e.g., X.25, etc., and routes the data to the ATM node where the medium access control (MAC) is located. At this point, the data is packaged and address routing performed, after which the data packets are sent via the SONET interface to the proper destination. The destination architecture is the reverse of that just described where the packets are decomposed in opposite manner.

The ATM network capability, like packet networks, can be located within the public switched network, or outside of it. There are 3 distinct possibilities for the association of the ATM network with the public network. These are represented in Figure 5-3.

5.1.3   Coaxial Cable

Coax, as coaxial cable is called, has the ability to carry a wide bandwidth of information. Coax has been used almost exclusively in community-access TV systems, known as CATV. The systems that have evolved around this technology, and its supporting electronics capabilities, are essentially 1-way rather than interactive, and have to be reamplified at relatively short intervals, because the attenuation associated with coax is significant.

The repeaters used in CATV systems only amplify in 1 direction for the most part, and therefore, signaling back to the servers and head-end equipment is not possible. Two-way capability has been achieved when 2-way repeaters have been installed in the transmission pathways; however, few cable ways have been so equipped because of the added expense involved. Even those systems in which 2-way repeaters have been installed have problems because of the additive noise buildup experienced in such circumstances. The tree structure configuration of CATV systems causes noise to be added as the signal flows back up the tree to the head end. Typically, all possible selections, i.e., the channel offerings, have to be presented to the termination points at the user end, at which point the user can make his/her own selection using some type of demodulator or channel changer.

5.1.4   Twisted-Pair Copper Wire

Recent advances in integrated circuitry, in particular those associated with VLSI and VHSIC chip manufacturing technologies, has made it possible to code, modulate, transmit, receive, demodulate, and decode digital signals of much higher bit rate than was previously the case. Essentially, the new signal processing technology has allowed the signal to be more corrupted in its received form and still be recoverable. This increased distortion is the result of higher data rates for the transmitted signal, thus necessitating tighter tolerances and shorter transitions in the modulated pulse trains. As these tighter spacings are corrupted by noise and other distorting effects during transmission, the recovery of the signal is made more difficult, because the receiving circuitry has a more difficult time determining what is or is not a valid digital signal after all

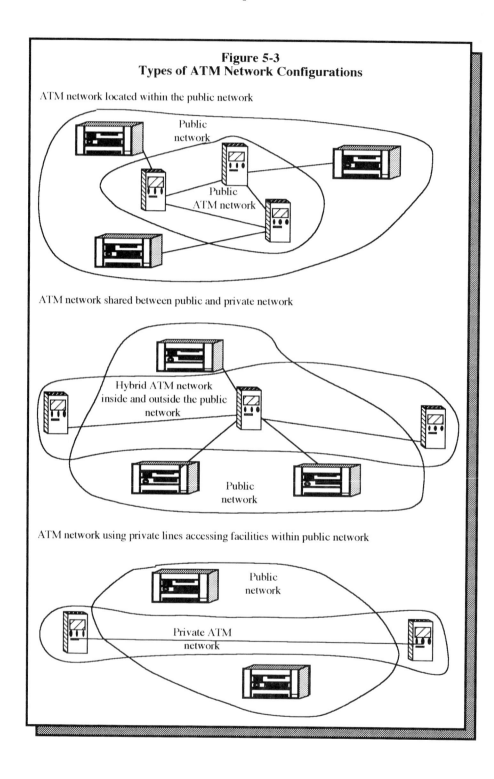

Figure 5-3
Types of ATM Network Configurations

ATM network located within the public network

ATM network shared between public and private network

ATM network using private lines accessing facilities within public network

receiving and decoding has been performed.

Various schemes have also been developed to support the transmission of analog information over copper twisted pair. These will be discussed later in this chapter. The general approach used for analog transmission is to boost certain frequencies and to shorten the path length in order to preserve the information content of the signal. Each boost and path length is referred to as a *transmission leg*. After each leg, the signal can again be boosted and retransmitted for as many times as the information can sustain such retransmissions. Ordinarily, the number of these retransmissions would be very few, probably no more than 2. All of these transmission capabilities have their own unique advantages and disadvantages.

5.1.5   Fiber Distributed Data Interface (FDDI)

FDDI is one of the current group of transmission technologies that is classified as a shared medium. This means that, unless fractional portions of the total link bandwidth are set aside for specific sessions, the total bandwidth of the medium is up for grabs for any demand that comes along and its use is dictated by instantaneous requests as they arrive at the FDDI interface. The problem, therefore, is one of providing a constant rate of information flow (bandwidth) for each request. For video and audio, this is difficult to guarantee, because each frame or burst may have to compete with other requests in the same network. As a result, there is no guarantee that a consistent video or audio frame rate can be maintained.

FDDI is a transmission and framing standard (layers 1 and 2) that has a nominal rate of 100 Mb/s without fractional allocations. The FDDI frames utilize a 48-bit address structure as defined by the IEEE 802.2 standard. There is a newer standard created to overcome the deficiencies of the original standard, called FDDI-II, that has the same data rate, but users can request specific bandwidth allocations on the network, such as 6 Mb on one or more links. In this way, a subchannel is defined for the user during the period of need. If each of 6 users want to reserve 6-Mb channels for video, then the remaining set of users, no matter how many that group may be, will have to be satisfied with 64 Mb of bandwidth for operational use. This situation is referred to as bandwidth-on-demand, because bandwidth requirements for each user are based upon the instantaneous need of each user at any point in time.

Because of its limitations, FDDI may be superseded by other technologies that rely upon greater bandwidth and information-laden packets. One such technology, discussed above, is ATM/SONET.

5.1.6   Frame Relay

Frame relay is a relatively recent technology concept. As a protocol, it encases the format of another protocol format, such as TCP, by providing standard header and trailer information that is not consistent with other formats. The theory is that even though the overhead is increased with extra header and trailer information, the throughput is also increased by simplified routing and administrative control. The basic frame relay architecture is shown in Figure 5-4.

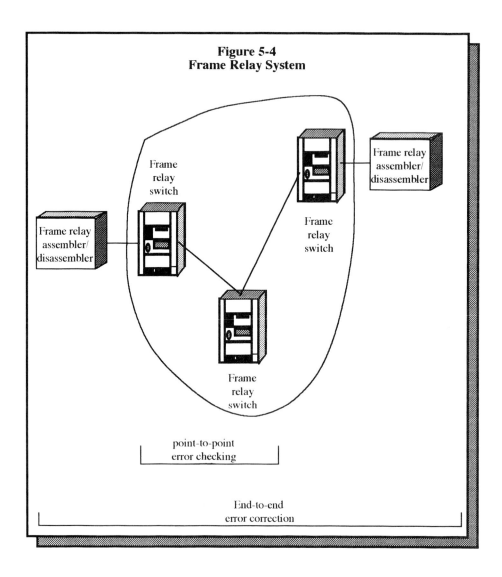

Figure 5-4
Frame Relay System

Unlike, for instance, the X.25 packet system, the frame relay network operates so as to perform error detection at relay each node and then discarding the packet(s) in error. Error correction is achieved after the destination node realizes that the packet(s) in error are missing by requesting retransmissions of those packets. The more sophisticated protocols, such as TCP/IP carry packet numbers, and therefore, the frame relay packet overhead has to be stripped before the missing packets can be identified. Frame relay networks employ variable length packets and incorporate flags for congestion and discard eligibility control, thus contributing to their routing and control efficiencies. The frame relay system is thought to be unsuitable for packet video and audio transmissions, because of its use of variable length packets.

5.2   Attributes of Transmission Systems

5.2.1   Demographic Considerations

There are over 200 million telephone lines installed in the U.S. and Canada. Practically all of these telephone subscriber lines in North America, for instance, those lines that carry voice between residences and businesses to other residences and businesses, pass through central (switching) office facilities. The central offices, in turn, are linked to each other via interoffice trunks that carry many circuits between these facilities. These distances may be many miles long, whereas the line distances between central offices and line terminations are less than 18,000 feet from their switching offices. In the 1920s, when the telephone system was being rapidly expanded, line noise was a major concern, because the equipment employed at that point was mostly mechanical. As a result, designers of the network decided that the analog lines must be filtered in order to assure some reasonable quality of the received signal. This filtering was referred to as *loading*. Loading means that the lines are intentionally filtered so that higher-frequency components are filtered out, or rejected.

5.2.2   Historical Considerations

This filtering implementation was originally instituted because the phone company facilities were so noisy that filtering was seen as the best method of reducing unwanted noise and improving the quality of the service. The reason why the loading of telephone lines was intentionally implemented by the telephone companies was because older technology connections and cross talk within the phone company facilities had very noisy elements that corrupted and compromised the quality of the information carried. The conclusion reached was that the elimination of noisy signals could be accomplished by filtering out these components rather than engineering smoother operating elements. The loading coils used today are set at about 3 kHz, thus allowing reasonable-quality voice utterances to be sent back and forth and identification of these utterances to be made.

Unfortunately, only higher and higher frequencies can support higher and higher data rates. For high-rate transmissions, above about 19.2 kHz, to be made, the lines must be unloaded, and in the place of the loading coils sophisticated modulation electronics must be employed to overcome the inherent noise of the system. In today's service environment, restricting the lines in the carrier serving area (CSA) to 3 kHz or less prevents high-data-rate services from being offered. The solution is to remove line filtering, to decrease noise sourcing through electronic switching, and to add sophisticated signaling and coding capabilities, in the form of chip technology or ASICs, to upgrade the data rate capabilities possible.

5.2.3   Technical Considerations

Additionally, however, there are other problems associated with the use of twisted-pair copper transmission for relatively wideband data rates. Among these are the switching noises associated with the typical telephone central office, the deplorable conditions of the copper wire, wire bundles employed between central offices, known as interoffice trunking, and the CSA infrastructure that supports the outside plant between the central office and the home. These problems introduce additional noise into the lines and trunks over which information is carried, such that the bandwidth and carrier ranges are limited. Notwithstanding these issues, multimedia/video carriage may be a viable possibility within the existing copper networks of the telephone companies if new technology approaches can be implemented.

This recognition gives way to the idea that video may even be carried over the existing phone lines if the appropriate technologies can be brought to bear upon the problem of bandwidth expansion. If video can be carried, then we can begin to think of multimedia applications as candidates for relay over interactive circuits originally conceived for voice only. There are 2 principal approaches to the carriage of video information, based upon the 2 current requirements, one being the video telephone and the other being the use of telephone lines as substitutes for coaxial cable used for the carriage of video signals. The first application requirement can be further refined as a need for the 2-way transmission of video signals, and the second requirement needs a 1-way video transmission path and a reverse path for channel switching signaling.

The next sections will discuss just these types of multimedia applications and options. A pictorial representation of the principal issues involved is given in Figure 5-5. This figure depicts the basic building blocks for the exchange of information between the session users. Currently, certain portions of the modal integration are receiving particular attention, including:

(1) Protocol packaging

(2) Protocol receiving

(3) Data routing

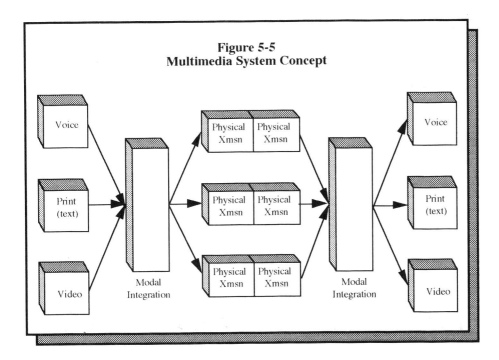

Figure 5-5
Multimedia System Concept

There are certain questions of packet sizing and overhead that will again be visited as the bandwidth availability and bandwidth needs are addressed. This will give way to variable-length packets for different network configurations or different modalities.

5.2.4   Economic Considerations

The existing copper plant used by the telephone companies in North America, and the physical connections to that infrastructure, are providing a very high percentage of the revenue for the telephone companies (telcos) and, based upon third-party predictions, will continue to provide that revenue well into the twenty-first century. In spite of the euphoria over fiber optic transmission linkage, fiber optic technology will not be ubiquitous for the next 20 to 40 years. Until then, copper will most likely be the mainstay of the physical communications offerings for the phone companies and their principal revenue stream. Added to this are the requirements that copper support ISDN primary rate access for T-1 rates. At the present, T-1 rates can be achieved on copper only at considerable expense and after appreciable time delay for the provisioning of T-1 service. The average installed first cost (IFC) estimate for a 1.5 Mb/s service varies between $1500 and $4000, depending upon the source of information used. This is a costly IFC for potential users and quite likely will remain a barrier to entry for many customers.

The challenge for the telcos, then, is to get the most out of the existing copper plant in the most cost-efficient manner while at the same time providing the services required for new offerings, such as ISDN, at prices that encourage service orders rather than discourage such orders. For the consumers, their costs will follow the cost reduction efforts of the telcos. Also, and in the meantime, third-party providers may be encouraged to file for permission to co-locate at central offices within the region for the purpose of offering new or existing services at lesser costs relative to the telephone companies. Given the Spartan labor rates and overhead structures exercised by some utility and communications companies, e.g., cable TV companies, their equivalent service offerings could be an important penetration of the telco business market.

5.3   In-Depth Analyses of Transmission Technologies

In this section, the reader will see discussed some technologies that are not often covered in telecommunications books [1:301-316]. The standard leading-edge transmission standards, such as ATM, are discussed earlier in this chapter or in other places in this text. As mentioned previously, copper will continue to be the dominant carrier of information for many years to come. As a result, we must be prepared to deal with that fact. Therefore, those technologies that can be used to extend the utility of copper must be understood, and these are presented here.

5.3.1   High-Bit-Rate Digital Subscriber Line (HDSL)

General Technology Issues

In order to address the various threats and problems cited above, and others like them, the telephone operating companies have to seek new cost reduction measures to cut the costs of installation and operation, especially for business and growth services. One of the emerging technologies that has been proposed and acted upon to solve this problem and to meet these challenges is called *high-rate data subscriber line*, or HDSL for short. This technology was originally investigated in order to meet IFC reduction objectives for ISDN. It incorporates several digital signal processing techniques and integrated circuit methods that have become available within the past few years. It will be explained more fully below, along with the problem behind its original investigation. Following these discussions, a review of the risks involved will be provided and the schedule for HDSL implementation will be proposed.

Architecture

The top-level architecture calls for 2 HDSL units to be used side by side, communicating with like units at distant locations within the *carrier serving area* (CSA), each transmitting and receiving an 800-Kb/s data stream, full duplex, for a total aggregate data rate of 1.5+ Mb/s in each direction. Each pair of HDSL units at each end of the local loop will utilize one twisted pair. T-1 data streams received at the central office or the remote distribution point will be subjected to a special coding and modulation regimen prior to transmission on the twisted pair to the customer premises. The overall architectural concept is illustrated in Figure 5-6. The HDSL implementation will conform to the CSA definitions and ANSI Standard TI.403-1989, which defines the network interface for T-1 service delivery to a customer.

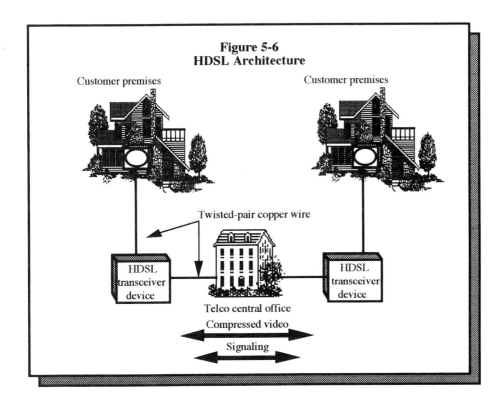

Figure 5-6
HDSL Architecture

Technical Feasibility

There are major technical issues that drive the realization of HDSL, these being the coding scheme to be used that provides some degree of compression, the modulation scheme that assures a satisfactory bit error rate, and the use of VLSI chip development that assures sufficient speed of processing for data handling. Technical Note 5-1 outlines the basis of the technology advantages. Each of these technologies will be discussed briefly below.

Several transmission codes have been investigated by Bellcore, university groups, and manufacturers. The final decision on which type of code to use favored a simple scheme known as the 2B1Q code that pairs the incoming bits such that 4 combinations are possible, namely, 00, 01, 10, and 11. These known combinations can now be used in the modulation scheme to represent a frequency or phase shift of some known type. The modulation scheme is referred to as frequency shift key, such that each shift in the wave shape of the signal represents a specific code [6:20-49].

The modulation scheme for transmission is an attempt to accommodate error requirements as well as data compression. At present the technique of choice is the quadrature amplitude modulation (QAM) scheme. Two carriers each 90 degrees out of phase with the other are used to compact the data flowing through the lines.

Technical Note 5-1
Very-Large-Scale Integrated Technology

Very-large-scale integration (VLSI) technologies have made it possible to take phase shift key (PSK) coding to new levels of granularity, or segmentation, and data stability. Traditional quadrature phase shift key (QPSK) circuits encoded 4 levels of information, in diatonic sequences, into 4 positions on the transmitted wave form. Thus, binary 00 would be encoded as a 45 degree shift. A binary 01 would become a 135 degree shift, a binary 10 would become a 225 degree shift, and, finally, a binary 11 would become a 315 degree shift.

VLSI technology takes this approach to another level of sophistication. The wave form can now be shifted at closer intervals along the wave form. A 16-point encoding scheme takes a run of 4 bits (strings of ones and zeros), and divides the wave into 16 segments for each of the possible values associated with a 4-bit string. Thus, a 360 degree sine wave is now shifted every 22.5 degrees in order to encode new digital values. Thus, a binary value of 0000 is recorded as 0 degrees. A 0001 is placed at 22.5 degrees, a 0010 at 45 degrees, and so on. This technology has been around for several years, and its widespread usage in this and other applications has been enhanced by the commercialization of VLSI technology, and soon VHSIC technology.

In computer processing, certain instructions consume much more time than others, and if the particular application calls for a high percentage of those high-time-consumption instructions to be executed, the application takes longer and may slow down other processes involved.   Larger and larger scale integration, such as VLSI, as opposed to LSI, provides the advantage of executing all instructions at a faster rate due to the higher density of packaging and hence the shorter distances for signals to flow.   The HDSL application addressed here has a considerable number of compare instructions that drive up the total execution time of the chip operation.   VLSI provides an additional time margin needed to assure that the chip will not be overloaded.   VLSI chip design and manufacturing are well-known processes that make this technical issue very manageable.

In spite of the ongoing pressures for the expansion of communications bandwidth through deployment of fiber optics, the fact remains that copper is, and will be, the principal method for telecommunications transmission via the land line method for the foreseeable future.   Given this scenario, improvements in the bandwidth of copper are a high priority for telecommunications providers.

Both of the high-density chip technologies, namely, VLSI and VHSIC, provide for closer spacing of the signal paths within the chip layers and, therefore, the chips can be made much smaller than those made with older technologies, such as LSI.   Additionally, these more compact packaging arrangements allow the signals to travel shorter distances and achieve faster interfaces with different components.   These shorter signal paths also make for more accurate results with lower error rates than is the case with older technologies.

The top-level architecture for HDSL calls for 2 VLSI units to be employed side by side, each communicating with like units at distant locations within the CSA.   HDSL operates by utilizing 2 sets of twisted copper lines.   One set carries digital video and voice information in one direction, while the other carries digital video and voice information in the opposite direction [5:4].   This facility is suited for digital video conferences, provided that the copper transport and switching equipment are properly conditioned.   Conditioning is a line condition where the signal transmission is unimpeded.   Conditioning requires that many problems associated with impedance mismatching, signal conditioning, and noise be overcome, among which are:

(1)   Cross talk between copper lines

(2)   Severe attenuation, or loss of signal strength over distances of 10,000 feet or more

(3)   Impulse noise generated by switching equipment and pickup from unshielded copper lines

(4)   Intersymbol interference where frequency components spread over time and overlap one another

(5)   Radio-frequency interference caused by the pickup of power equipment and radio transmitter energies

Even eliminating these problems makes the HDSL technology limited in application at this point. Aside from this, however, it is specified to operate over distances of between 9,000 and 12,000 feet and at T-1 rates of 1.544 Mb/s in each direction. Meeting these expectations is a challenge that may be irrelevant as certain wireless techniques become available.

Each twisted pair operates by transmitting and receiving an approximately 772-Kb/s data stream, full duplex, for a total aggregate data rate of 1.544 Mb/s in each direction. Each pair of HDSL units opposing each other at either end of the local loop will utilize one twisted pair. T-1 data streams received at the central office or the remote distribution point will be subjected to a special coding and modulation regimen prior to transmission on the twisted pair to the customer premises in order to assure the accurate reception of the signal at the receiving end.

Equipment Availability

At the present time there are several companies that have developed or are working on HDSL transceivers. The vendor community has been very active in the T1E1.4 committee to get its needs met for the anticipated sales of this technology.

5.3.2   Asymmetrical Digital Subscriber Line (ADSL)

Somewhat closely akin to the HDSL technology is another approach known as ADSL. The target market is different from HDSL, but its technique of implementation is about the same.

ADSL is based upon the use of one twisted pair of copper wires to deliver a composite video and sound picture in one direction, and reverse channel signaling in the opposite direction over the same line. The total bandwidth is also a T-1 rate of 1.544 Mb/s. For both HDSL and ADSL, a standard video rate of 30 frames per second and quality picture are the objectives. A normal digital, 30 frames per second, NTSC standard picture normally requires about a 45-Mb/s transmission rate. Obviously, some very sophisticated coding and compression has to be performed in order to reduce this requirement to that of a T-1 specification, i.e., something close to 30:1 for a 1.544-Mb/s transport rate. This architectural concept is illustrated in Figure 5-7A.

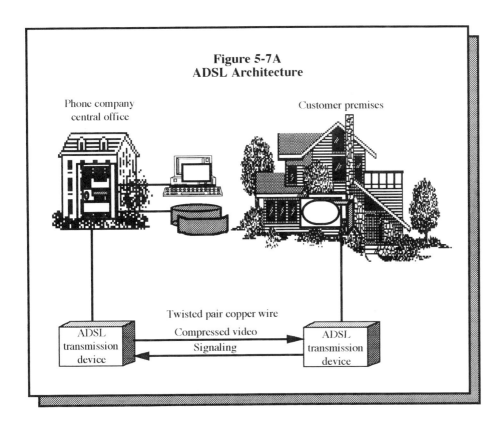

**Figure 5-7A
ADSL Architecture**

Phone company
central office

Customer premises

Twisted pair copper wire

Compressed video

Signaling

ADSL
transmission
device

ADSL
transmission
device

Both the HDSL and ADSL technologies are implemented via interface units, or "boxes," at both ends of the physical narrowband link. These interface units provide the encoding and decoding of the modulated and demodulated signals entering at either end of the link. ADSL was designed to operate over distances of 12,000 to 18,000 feet, or about 5490 meters. HDSL was designed for distances up to 12,000 feet, or about 3660 meters. ADSL has been considered as a method for video programming transport, and it may be suitable for certain types of programs. However, the selection of channels or sources must, with ADSL, be provided from some remote point further up the transport route than the ADSL designated distances.

The generic outline of the ADSL technology is depicted in Figure 5-7B. The video is compressed initially to narrow the transmit bandwidth requirement. This compression is a multistep process whereby the image pixels are, first, encoded so as to reflect the most efficient means of transmitting the image information possible. Next, some variant of phase shift coding is employed to pack the already compressed information into the least possible transmit bandwidth on the carrier wave. In terms of time, this is most efficiently accomplished by using a special VLSI chip, sometimes called an ASIC (application specific integrated circuit) that codes the digital signal according to coding schemes embedded in the chip. Following this, the modulation is performed by applying an analog representation of the information on to the carrier frequency. This final compression step is filtered to assure as clean a signal as possible before output to the copper lines [3].

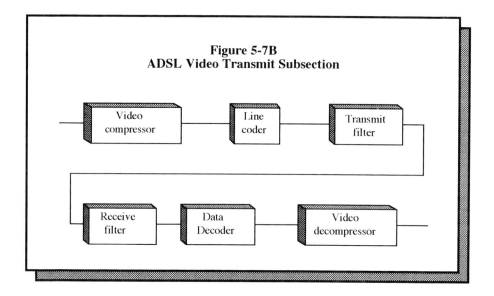

Figure 5-7B
ADSL Video Transmit Subsection

At the receive end, the receive filter conditions the signal, by again filtering the incoming information to remove noise, and then converts it from an analog to a digital signal representation. The signal is decoded so that the transmitted digital stream is recovered, and finally, the recovered signal is decompressed so that the extended bandwidth is obtained.

The ADSL system interface is the digital signal stream that is to be encoded upon the carrier in one of several varieties of phase shift key transmission modalities. Likewise, at the receiving end, the decoding step passes a digital signal to the video decompression device at which point the pixel decompression occurs. Thus, the compression occurs at 2 levels, one of which is within the ADSL technology, and the other of which is outside this technology. ADSL can provide up to a 6:1 compression advantage depending upon many factors, such as the transmission line quality.

5.3.3   Analog Transmission Systems (ATS)

The ADSL and HDSL solutions to the transmission problem of video are digital ones. At least one analog solution to the same problem of video transmission over narrowband links, such as twisted-pair copper wire, exist, that we will refer to as analog transmission systems (ATS). This capability is marketed by several companies, such as Lightwave Systems. The idea of using copper lines for the transmission of video originated with the U.S. Secret Service's concern over monitoring the grounds around Camp David, where high-level presidential conferences and relaxation activities occur. The basic concept for this technology approach is illustrated in Figure 5-8. These systems have the following advantages.

(1)   They are simple to implement. There are no additional telephone company services that need to be purchased in order to realize the benefits of video over copper. The interface units cost about $1100 in small quantity pairs. In large quantities, these units are about $400 per pair.

(2)   The units come in different operational capability configurations, i.e., 1-way transmission only as well as 2-way versions, where there is a simultaneous downstream video delivery transmission and an upstream signaling capability.

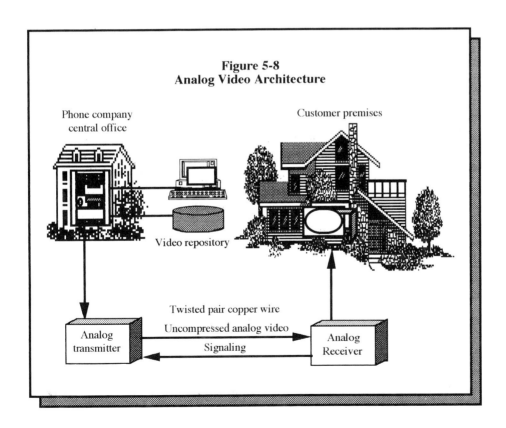

Figure 5-8
Analog Video Architecture

(3)   The 2-way delivery and signaling systems have the ability to be
       used in a fashion analogous to the familiar community-access TV
       system (CATV), where the user can use a wand to request channel
       changes. The difference between this system and the CATV
       system is that the ATS system signaling requests are relayed back
       to a switching unit some distance away, such as 5000 feet, where
       the downstream signal is switched from the one currently being
       viewed to the requested channel. The CATV system carries all of
       its offerings to the viewer simultaneously, and channel changing is
       accomplished at the set, as opposed to at the server. This
       approach has the reward of being able to carry an indeterminate
       number of channels.

The disadvantages of this system approach are as follows.

(1)   The carrier distance is limited to about 4000 feet, or 8000 feet with
       an additional stage of amplification.

(2)   A receiver unit must be located at each termination end and a
       similar transmission unit at the sending end.

(3)   The present trends toward digital transmission schemes may make
       the analog solution hard to sell.

There is no question that analog data streams will have a limited appeal
as digital technologies become more cost-effective and widely used. However,
their utility is as an adjunct to digital transmission systems and a temporary
solution in many cases until fully equipped digital systems are available in
certain areas.

Summary

Multimedia transmission systems are divided into 2 major categories, one being broadband systems, such as ATM switches coupled to SONET transmission links where true multimodal and interactive exchanges of voice, print, and video can take place, and the other being narrowband technologies that are combined with compression and multiplexing schemes to achieve the bandwidths needed. The problem with these narrowband solutions is that they are inherently limited in their utility as demands for bandwidth increase. The problem with broadband solutions is their complexity and recent innovation - many bugs will have to be worked out in order to achieve the results desired. For example, regarding ATM transmission speeds in some interface systems where translation from ATM to ethernet is occurring, the nominal speed of 155 Mb/s is desired but in reality 100 Mb/s can only be achieved. Thus, much work will have to be done in this area to meet expectations. The technologies developed for today's broadband needs include limited analog techniques as well.

Chapter Five

<u>References</u>

(1)   "An Overview of Cable and Wire Technology," *Datapro Reports on Telecommunications*, McGraw-Hill, New York, 1990.

(2)   Goldman, J. E., *Applied Data Communications*, John Wiley, New York, 1995, p. 516.

(3)   Gonzalez, P. Wintz, *Digital Image Processing*, Addison-Wesley, Reading, MA, 1979, pp. 5-227.

(4)   Held, G., *Data Compression*, John Wiley, New York, 1983, pp. 18-57, 72-90.

(5)   Hsing, R., J. W. Lechleider, and D. L. Waring, "HDSL and ADSL: Giving New Life to Copper," in *Bellcore Exchange*, Livingston, NJ, March/April 1992.

(6)   Raymond, V., *An Introduction to Source Coding*, Prentice Hall, New York, 1993.

Chapter

6

Multimedia Technologies: Interfacing with Network Systems

Chapter Highlights:

Over the years, many of the large or influential telecommunications services providers have developed and marketed their own versions of various types of network management and surveillance tools. Because of their own strategies and product backgrounds, their various management systems have vastly different objectives, capabilities, and capacities. The leaders of the network management efforts include AT&T, IBM, GTE, and Novell, to name a few. This chapter will discuss in detail, the architectural approaches of these industry leaders, and illustrate these approaches with representative screens and/or diagrammatic layouts.

6.1   Existing Network Systems for Multimedia

Each manufacturer of communications equipment, such as T-1 multiplexers or modems, also provides a network management capability for that element type. Aside from the subnetwork management systems that exist for each type of network element, such as modems, packet assemblers/disassemblers, T-1 multiplexers, etc., several total network management architectures have been successfully developed and marketed, intended for the management of heterogeneous systems. Unfortunately, none of these facilities nor those that purport to manage homogeneous components, such as T-1 multiplexers, currently have the capabilities to monitor or control the activities needed for multimedia network requirements [9:20-45].

The 2 principal heterogeneous network management systems that have been developed include AT&T's Unified Network Management Architecture (UNMA), and IBM's Netview and Netview PC, which is IBM's companion product for non-IBM equipment. In addition, others, such as HP and GTE, have also pursued their own views of network management, and Novell has successfully become the *de facto* network management standard for LANs, with about 60 percent of the market. Unfortunately, none of the various approaches to network management was originally developed with the coming era of multimedia in mind. However, it appears that most vendors in the data and multimedia arenas have made allowances to accommodate multimedia requirements. Some will have more success than others in modifying their product lines to serve the peculiar needs of multimedia. Each of the major architectures and their approaches is discussed below.

6.2   AT&T's Unified Management Architecture (UNMA)

6.2.1   Background

AT&T has approached the problems of network management with the idea of providing a total integrated management environment for both the public network elements, such as HDSL, and fractional T-1, as well as the private elements, such as modems, PADs, etc. This philosophy is based upon the premise that its biggest customers from a dollar-volume point of view will require a wide-area geographic solution. It turns out that the largest customers in any of the telco areas, whether they be local exchange providers or interexchange carriers, account to less than 20 percent of the total customer base, but provide more than 80 percent of the revenue [10:1-10]. Thus, the AT&T supposition is probably a correct one. With this premise in mind, AT&T has elaborated its concept to include the following objectives and constraints:

(1)   Offer a system that embodies end-to-end network identification and the integrated management of an entire and heterogeneous array of network elements.

(2)   Develop a network management system that follows as much of the ISO standards as apply to the entire 7 layers of the network management schema, but which focuses on layers 1-5, i.e., the physical layer up to and including the session layer.

(3)   Design and develop a system that employs a distributed processing design which would provide global capability.

(4)   Emphasize the management of transmission systems with initial development focus upon fault management facilities.

(5)   Support the development of standard interfaces for multivendor environments. Such interfaces would be data exchange interfaces as well as user interfaces.

(6)   Assure that there is adequate separation of the applications modules from the network management functionalities.

(7)   Emphasize the design and implementation of machine-to-machine interfaces for its network management offerings.

(8)   Develop management strategies based upon a decentralized approach such that regional and local control and information can be made available to the targeted customer population, and local control over fault or operational problems can be exercised in the shortest possible time frame.

All of these ideas have been incorporated into the generalized AT&T network approach that is illustrated in Figure 6-1. Under this concept, AT&T conceives of the network management problem as one of end-to-end integration of all aspects of the communication problem, in particular the physical components of the transmission linkages. For AT&T, the interexchange carrier segment of the architecture is easily understood since AT&T is an interexchange carrier.

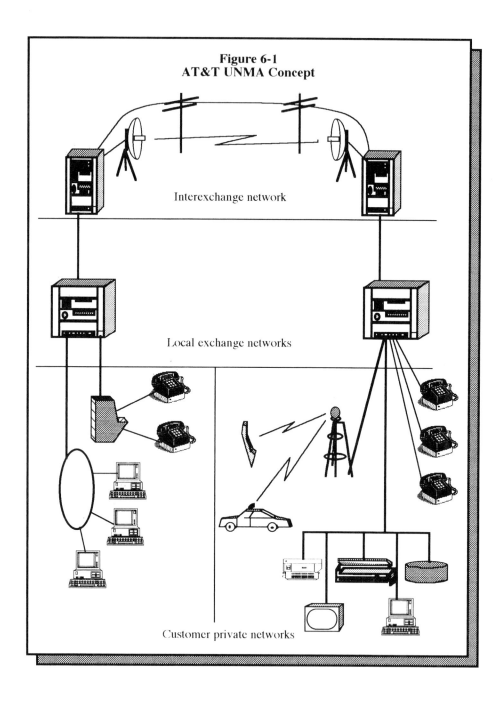

Figure 6-1
AT&T UNMA Concept

Interexchange network

Local exchange networks

Customer private networks

Presumably, AT&T can provide expert management of all pertinent issues associated with the first 3 layers of the ISO model since these are addressed by the AT&T network directly. Transport and sessions (layers 4 and 5) can be additionally addressed as an extension of the AT&T infrastructure. The local exchange carriers, with whom AT&T interfaces, are assumed to provide circuit status information to AT&T's network manager for those circuits being tracked by the AT&T system. Finally, the customer network will supply AT&T with data concerning each of its local installations. These local networks are supported by AT&T's points of presence (POPs), which also allow customers, especially large customers, to bypass the local carriers directly into the AT&T network, thus minimizing the impact of any missing information about the customer's network operational status. Figure 6-2 shows the elements of the network shown in Figure 6-1 but with the additional features of the localized network managers included.

Typically, vendors of the similar network elements, such as packet network equipment, T-1 multiplexers, modems, and the like provide managers with their own equipment sets. The AT&T approach also incorporates and accommodates element managers that provide network surveillance over equipment mixes that are within the same geographical or functional areas.

6.2.2   Multimedia Technology Impacts

The implications of the AT&T approach to network management are significant insofar as multimedia management is concerned. Recognition that all 7 layers of the ISO model are important means that the applications and presentation issues will eventually be addressed by the UNMA architecture as the requirements build and become better defined. As discussed in previous chapters, data networks either do or do not get the intended information from point A to point B, based in large part upon the bandwidth of the carrier facilities and upon the packaging of the information transported.

In the case of multimedia information, whether the information gets through is only part of the problem. The resulting quality of that information is the more crucial issue. Since AT&T is committed to an eventual total end-to-end solution encompassing all 7 layers of the model, the AT&T approach may be well positioned to assure media quality when all pieces of the UNMA are put into place. As applications management becomes more and more differentiated from lower-level OSI layers, the AT&T system will keep pace with such development activities because of its prior focus upon wide area communications management and end-to-end connectivity.

The emphasis upon end-to-end connectivity is another point that potentially makes the AT&T network management architecture appropriate for multimedia applications. One of the conditions that qualifies the usage of quality information in a multimedia network is the problem of how that information is to be conveyed from the originator to the receiver. This process may be more difficult if the network manager is deficient in its ability to assure end-to-end oversight. The remaining technologies are those that assure bandwidth availability and information packaging.

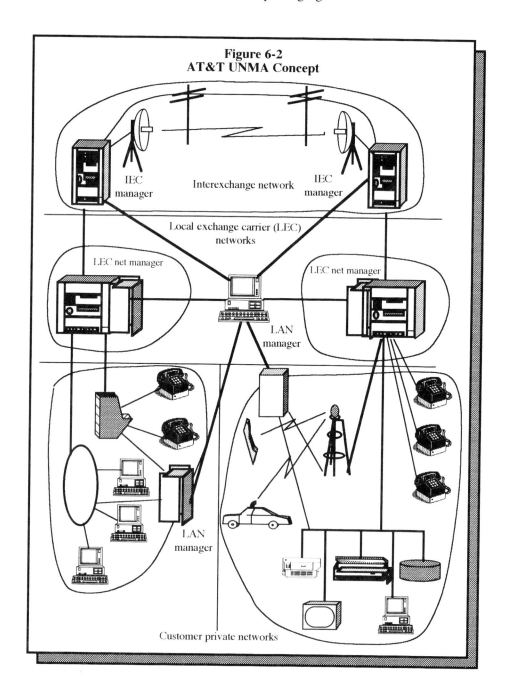

Figure 6-2
AT&T UNMA Concept

6.3   IBM's Netview

6.3.1   Background

IBM's view of the network management environment is somewhat different from that of AT&T. IBM grew up as a large mainframe computer company. For this reason, its focus has always been upon computer networks, in particular hierarchial systems where there is 1 mainframe to which many peripheral devices are attached and upon which they depend for systems and applications support. IBM has created a Systems Network Architecture (SNA) that is its cornerstone for network management and control. SNA requires that certain protocols be exercised during device communications. Non-SNA devices do not recognize these protocols, and therefore network management of these devices is rendered incapable. The IBM approach to network management is summarized below.

(1)   Focus should be placed upon host-based systems, most of which are seen to be IBM or SNA-compatible. Smaller networks, such as LANs, were originally ignored, but were considered later mainly because they are based upon distributed processing implemented on workstations.

(2)   Communication between network elements consists primarily of data, while the management of other modalities, such as voice and video, will result from the integration of data with these and other communications media.

(3)   Carrier networks are to be considered as a series of extended interfaces between various local networks, both SNA and non-SNA.

(4)   Maintenance should be conducted in a pro-active manner and conducted continuously as opposed to when customer complaints occur or during routine testing sessions of that equipment.

(5)   The objectives of network management will best be served through a centralized network manager most likely to be implemented on a mainframe machine. Decentralized network management requires additional coordination between machines

which appears unnecessary in light of the power of the primary mainframe.

(6)   The packet orientation of the various broadband technologies lend themselves to analyses on a circuit basis because of the upper-layer monitoring possible by the IBM architectures.

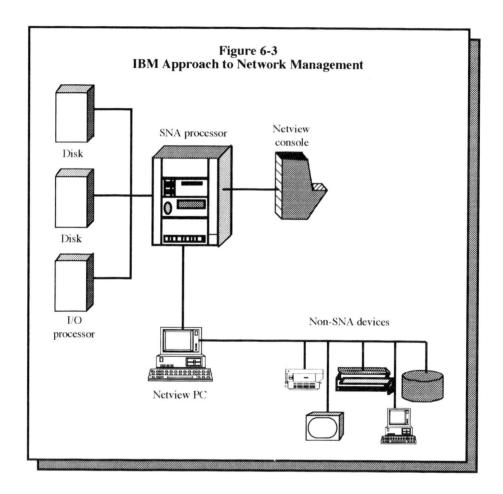

Figure 6-3
IBM Approach to Network Management

IBM is a long-time manufacturer of computer equipment. During the 1960s, the usage of computer systems began to change. Local attachment of terminals used to control computer mainframes gave way to numerous remotely attached terminals mutually connected over telephone lines. Thus began IBM's dual role as processing equipment supplier and communications provider. IBM's interests still resided with selling computer equipment and peripherals, but remote access to computers made it necessary for IBM's interests to expand to include modems, multiplexers, and the like. The IBM approach to network management can best be explained by referring to the concept portrayed in Figure 6-3.

IBM defined a set of facilities for network management under its Systems Network Architecture (SNA) philosophy. This capability was named Netview, and initially it was developed for its own array of equipment offerings. The network management environment, as IBM saw it, consisted of a multivendor environment that contains both SNA and non-SNA equipment and components. For those elements that were SNA compatible, Netview was sufficient for network management purposes, but non-SNA equipment items could not be properly managed without costly, device-specific software and hardware changes in order for the equipment interfaces to be of use.

Netview, as a network management facility, was developed by IBM to provide surveillance and control of the IBM equipment components that adhere to the Systems Network Architecture. Thus, Netview operates upon IBM mainframe systems. Soon after Netview was conceived, it became apparent to IBM that there were many equipment items, especially communications components, that did not subscribe to the IBM SNA protocol interfaces. Thus, either IBM would have to forget about the network management of non-IBM systems, or some other ancillary facility would have to be offered if IBM wanted to monitor the equipment items of other vendors.

IBM decided that ignoring non-SNA equipment items would not be in its best interests economically, as it was trying to create the image of having a universal architecture for all communications devices. Therefore, an add-on to Netview would have to be defined and implemented to support the increasing numbers of non-SNA vendors supplying equipment to the telecommunications industry. Originally, IBM pursued the concept of Netview PC where the non-SNA devices would be attached through the PC to the mainframe for monitoring and control. The principal configuration for the Netview PC environment is shown in Figure 6-4.

6.3.2   Multimedia Technology Impacts

The IBM approach to network management has emphasized the monitoring and control of data networks throughout the evolution of computer networking. As a result, IBM has concentrated upon the design and implementation of data networks as opposed to focusing upon bandwidth or transmission protocol issues. Also, the traditional IBM architecture has been one of centralized host services providing server nodes with the access for individual terminals and/or workstations. As a result, distributed networks have not been as much a part of IBM's approach to the solution for network problems as other architectures have tended to pursue.

Data networks are more concerned with whether or not the information was received in error, and thus, the lack of emphasis upon media issues is easily understood. Multimedia networks, as discussed in previous chapters, are more concerned with the quality of the received information. The typical data networks are also constrained by the bandwidth of the existing physical transmission media. These media have consisted mainly of twisted-pair copper wire, coaxial cable, microwave for long-distance carriage, and occasional use of fiber optics where direct interfacing is needed for bandwidth considerations. However, direct fiber interfaces with data networks are only now being implemented with ATM technologies.

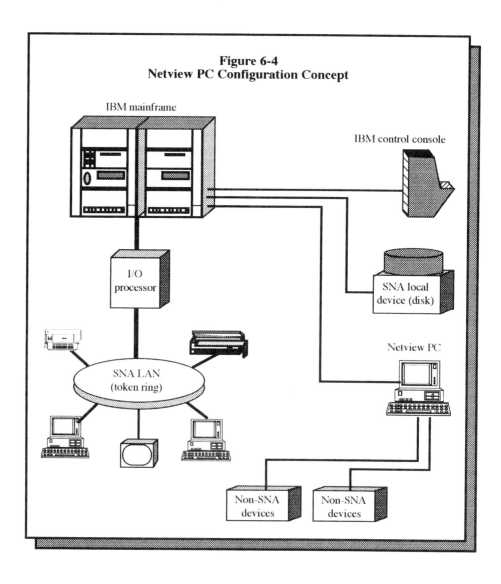

Figure 6-4
Netview PC Configuration Concept

ATM combines the transmission needs of various network elements with the flexibility for adjusting the bandwidth as the needs dictate. ATM is a technology that can provide a shell within which other data protocols, such as TCP/IP and X.25, can reside and operate. Thus, a vendor that provides communications tools and facilities is not necessarily positioned to support multimedia applications, because IBM is positioned in the upper layers of the OSI model, its software facilities.

6.4   GTE's Network Performance Monitor

6.4.1   Background

Local exchange carriers, such as the former "Baby Bells" of the AT&T family of companies, are focused upon the performance of their networks, and as a result, this concern becomes the primary focus for their brand of network management. GTE is also a local carrier, even though it is an independent telephone company [11:1-10]. It has developed a network management tool known as the Network Management Command and Control System (NMCC). This system performs several tasks, among which are:

(1)   The NMCC establishes the criteria for network performance operations by monitoring the actual operational status and comparing it against the theoretical operational capacities of the system monitored.

(2)   The NMCC collects and processes network data for use in the measuring of the performance criteria used for decision making.

(3)   The NMCC determines the degradation of services as they are affected by external circumstances, such as operational loading or configuration inadequacies.

(4)   The NMCC corrects network performance as conditions warrant by shifting resources or rerouting traffic as required to optimize operational needs.

The NMCC operates upon information stored in a database that defines the network configuration. The network configuration database consists of the types of information illustrated in Technical Note 6-1. The NMCC then analyzes equipment status and traffic conditions. This may be accomplished by measuring the differences between transmission and equipment capacities and their actual usage ratings. The database, which reflects the network status, stores information principally about the network traffic and alarm status. Software status is collected and indications are generated from thresholds and limits established by the users.

Technical Note 6-1
Typical Data Configuration Record for a Network Element

The typical configuration data record for a network element is as follows:

name of element
element serial number
element location
physical interfaces for element (1-n)
element capacities
element protocols supported
nominal element MTBF (mean time between failures)
nominal element MTTR (mean time to repair)

External alarms and operational traffic data are compared to the database record entries and their associated software comparison criteria. Alarms and traffic data that are outside the range of the software thresholds and limits are brought to the attention of the network manager. Such indications can be presented to the user in any of several different ways, but the most descriptive is via graphical icon display. The most elegant and useful solution to a problem is usually the simplest one. This is the case with the GTE system.

The software database definition for an NMCC alarm specification is indicated to the database as shown in Figure 6-5. The system developed is a natural-user interface style. The system faults are defined and specified to the system through a menu-driven interface. Traffic operations are monitored by a similar interface menu as shown in Figure 6-6. The range of faults indicated to the system manager include carrier failures, no call completions in a trunk group, and other such operational problems. Using these 2 types of specifications, i.e., system faults and operational outages, the local exchange carrier can monitor certain aspects of the entire communications picture.

6.4.2   Multimedia Technology Impacts

This type of system can be applied to the various transmission services available today and modified to accommodate the broadband transmission systems being deployed. The problem with this type of operational system monitor facility is that it is almost exclusively oriented towards the physical layer of the OSI model. Newer technologies, such as frame relay, ATM, and the like, are packet-oriented and as a result, the efficiency, and in particular the quality of the multimedia data, cannot be properly assessed without further monitoring of the upper layers of the OSI model.

Figure 6-5
Software Fault Definition

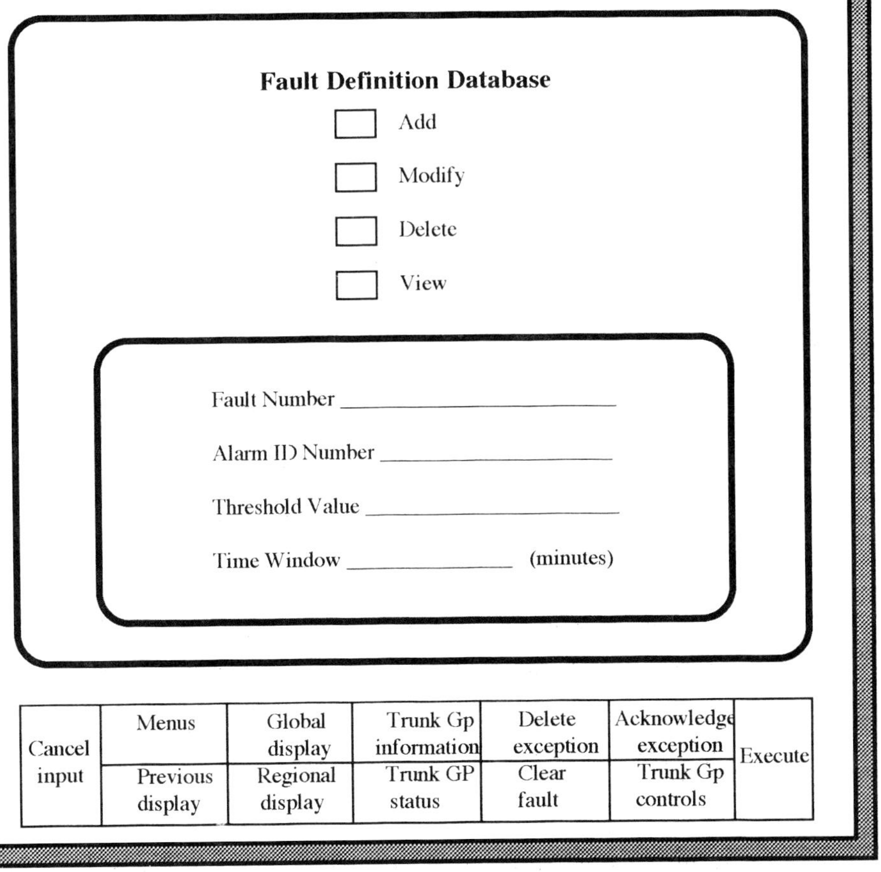

Fault Definition Database

☐ Add

☐ Modify

☐ Delete

☐ View

Fault Number _____

Alarm ID Number _____

Threshold Value _____

Time Window _____ (minutes)

Cancel input	Menus	Global display	Trunk Gp information	Delete exception	Acknowledge exception	Execute
	Previous display	Regional display	Trunk GP status	Clear fault	Trunk Gp controls	

For situations where the multimedia conversations are dedicated to individual circuits, the monitoring of these media interactions can be made on a circuit-by-circuit basis. However, where the media interactions are made on a shared broadband packet media basis, the monitoring of each virtual circuit (not in actual fact) is difficult on a physical layer basis. This situation is best addressed by monitoring each multiplexed circuit on the broadband link.

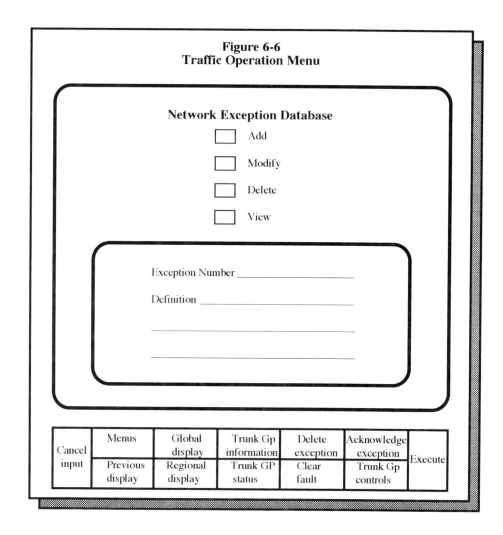

Figure 6-6
Traffic Operation Menu

Network Exception Database

Add

Modify

Delete

View

Exception Number _____

Definition _____

_____

_____

Cancel input	Menus	Global display	Trunk Gp information	Delete exception	Acknowledge exception	Execute
	Previous display	Regional display	Trunk GP status	Clear fault	Trunk Gp controls	

6.5   Novell's Netware

6.5.1   Background

Novell is one of the leading providers of local area network (LAN) equipment. The company originally designed and developed LANs and LAN management for the token ring architecture pioneered and commercialized by IBM. Later Novell expanded its LAN coverage to include ethernet. Its product line includes a series of hardware and software items, such as its network operating system (NOS), known as Netware, its LAN analyzer, known as LANalyzer, and its LAN manager, known as the Novell LAN manager. It also has the capability to interface with the Simplified Network Management Protocol (SNMP). Similar facilities have been developed by the other major vendors of LAN services [12:1-10]. For example, Microsoft has its own version of a LAN Manager, Banyan has its VINES operating system, IBM has its LAN Server, and Apple has its Appleshare system.

All of the facilities for Novell's LAN network operations were designed to be implemented within the 3rd, 4th, and 7th layers of the OSI model. Its interface with the SNMP also operates within the 3rd and 4th layers of the OSI model. At the applications layer, i.e., the 7th level, Novell provides its Novell core protocol (NCP) and service advertising protocols (SAP). For purposes of discussion, Figure 6-7 will serve as a guide for further elaborations upon the architecture and techniques for the monitoring and control of the Novel LANs.

Any LAN consists of a physical communications backbone plus the workstations, servers (file and information storage), and any other special input or output equipment, such as document scanners, printers, bar code readers, video players or video collection devices, CD-ROMs, etc. All of these components interact with one another according to a protocol defined and orchestrated by the LAN manager. Relating to Figure 6-7, we will begin discussing the entire system by focusing first upon the workstation.

All workstations must carry out specific applications according to the business or tasks that depend upon the platform and its network. The programs and data files needed for the tasks may be either resident internally within the platform or these sets of data must be retrieved from a server attached to the LAN or available from LANs connected to the user's LAN.

The workstation configuration shows that external interactions to servers, either locally or remotely, occur via an interface that begins with a Netware operating system shell known as the Network Client Extension (NETX).   This shell provides the interface to either the user interface environment or the network environment. Requests for data or programs to the network proceed through the bottom 3 layers of the OSI model. The NETX shell passes requests to a 3rd-level interface known as the Internetwork Packet Exchange (IPX). At this point, the request is packaged as a formatted message containing the sender's address, the destination address, the data request, and any other overhead information, such as error detection and correction coding, etc. The data packet is then passed to the 2nd layer where framing is performed, and then output to the physical transmission system.

The data packet is received at the server and proceeds back up the OSI layers to the IPX, and then the request is stripped from its format and passed to the network file server application. Requests to other workstations in the network are handled the same way except for the fact that the received request is forwarded to the NETX shell for resolution before the query to the local file system is placed.

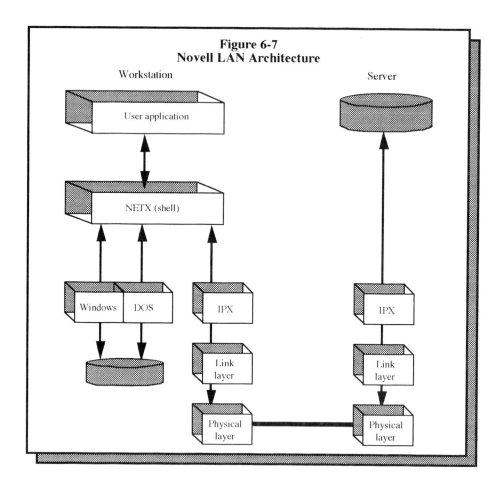

Figure 6-7
Novell LAN Architecture

There are several packet communications protocols that can be used for client-to-server and client-to-client interfaces. Because of the fact that not all workstations may be capable of decoding different protocols, these protocol handling and conversion features must be provided either at the workstation level or at some intermediate point so that nodes within a network, e.g., ethernet or token ring, may be able to communicate with one another. For protocol conversion and interpretation at the workstation, this must begin at the 3rd layer, where network addressing and packet formatting occur. In the diagram in Figure 6-7, this would occur at the point where the IPX is operating, and such protocol conversion features would operate alongside the IPX software under the name of NLM, discussed below. Each protocol conversion task would be initiated at the 3rd layer, with each type of protocol capable of being analyzed at that workstation being available for that task.

Novell has also developed a set of network extensions, along the Apple architectural model, that provide additional services for Novell users. These are called Netware Loadable Modules (NLMs) that are loaded on the servers of the network. These additional functional features include capabilities to recognize, decode, and interpret Appletalk, TCP/IP, and other OSI protocols. Other NLMs support different file systems, such as the Network File System (NFS) used commonly by UNIX clients, the Macintosh File System (MFS), and the File Transfer Access Management (FTAM) standard employed by clients using the OSI protocols.

The Novell NLMs also include a series of network management tools, among which is the LANTERN Services Manager that manages SNMP tasks. It is a network extension that operates in the background at the applications level and provides network status and alarms as required when requested by the operator or when dictated by the nature of the alarm.

Figure 6-8 illustrates the configuration options for the 2 principal networking protocols. Hubs are used by Novell to interface workstations, servers, and special input/output equipment to each other as dictated by the needs of the specific LAN requirement. In both cases shown here, the hub provides that connection interface facility.

6.5.2   Multimedia Technology Impacts

The current discussion of PC and Macintosh LANs is inserted here because multimedia users may well have to start with and utilize LANs for some time before multimedia capabilities are made available on wide area networks (WANs), metropolitan area networks (MANs), or across long-haul fiber optic network systems. Some LANs are fully capable of addressing and transporting video at 20 or more frames per second, thus indicating that multimedia applications can be handled quite adequately on such networks. The current problem with LANs is that they were not originally designed for the high data rates demanded of video and multimedia transfers, whether they be continuous, packet-formatted, or embedded in file structures. The problems that may affect the ability of current LANs to convey information include:

(1)   The protocols that current LAN technologies support were not designed to optimize bandwidth for the packet video and packet audio content on those systems.

(2)   The current LANs for the most part suffer from deficient data rate capabilities.

(3)   The current LANs lack effective management tools with which they could be managed.

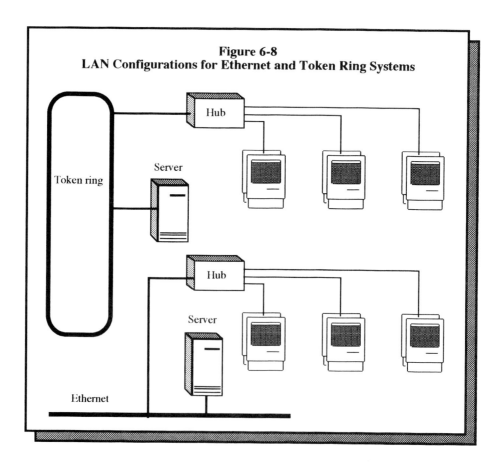

Figure 6-8
LAN Configurations for Ethernet and Token Ring Systems

Typical network management tools, as they are used today, do not discriminate between different modalities, but rather they discriminate among devices and protocols. The traffic information collected in the current models of management software identifies nodes and the data flows into and out of these nodes, but little else. The internals of what each data set contains in terms of data types, quality of information, and other pertinent parameters could lend additional and valuable indications of what is happening at each node. What is needed is a capability to look at each node and to identify its types of interactions, their quality status, whether the data link can adequately support the data rates required, etc.

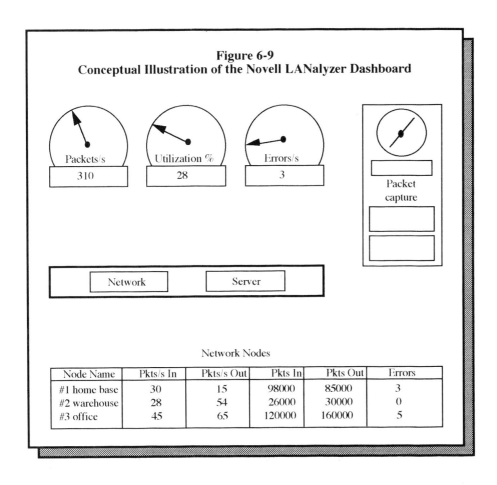

Figure 6-9
Conceptual Illustration of the Novell LANalyzer Dashboard

Network Nodes

Node Name	Pkts/s In	Pkts/s Out	Pkts In	Pkts Out	Errors
#1 home base	30	15	98000	85000	3
#2 warehouse	28	54	26000	30000	0
#3 office	45	65	120000	160000	5

The typical network management systems, such as that typified by the Novell Netware LAN manager, portray information about the physical and transmission status of the information carried by those linkages. An example of this data-type approach to management is shown in a stylized illustration of the Netware "Dashboard," which is the name for its system status overview. Figure 6-9 shows this status display for a conceptualized Netware LANalyzer. This type of display provides the viewer with a snapshot of the condition of each station in terms of the packet rates into and out of each node, the absolute numbers of packets into and out of each node, and the errors associated with each node. Also, an overall snapshot of the entire local area network is provided toward the top of the display in VU meter format, where the packets per second, system utilization is concerned, and total errors are accumulated and enumerated.

Summary

This chapter discussed and illustrated the various popular vendor configurations that can be used to implement a telecommunications network. Arguably, these same configurations could also support a truly interactive system needed to provide multimedia capabilities. The configurations discussed range from LANs, such as those provided by Novell, to the IBM architecture. GTE and AT&T both provide networking configurations based upon telecommunications systems components.

References

(1)   McGrew, P., *On-Line Text Management: Hypertext and Other Techniques*, Intertext Publications, McGraw-Hill, New York, 1989, pp. 40-50.

(2)   Goyal, S. K., and R. W. Worrest, "Expert System Applications to Network Management," in *Expert System Applications to Telecommunications*, J. Liebowitz (ed.), John Wiley, New York, 1988, pp. 1-150.

(3)   Harris, C. J., and I. White (eds.), *Advances in Command, Control and Communication Systems*, Peter Peregrinus, London, 1987.

(4)   Henning, W., "Bus Systems," *Sensors and Actuators A*, 25-27 (1991), pp. 109-113.

(5)   Hoffman, M., "Technology Profile Neural Networks," *Techmonitoring*, SRI International, July 1991.

(6)   Lemmon, A., *Marvel - A Knowledge-Based Planning System*, GTE Laboratories, Internal Report, Waltham, MA, 1986.

(7)   Martin, J., *Design and Strategy for Distributed Data Processing*, Prentice Hall, Englewood Cliffs, NJ, 1981.

(8)   Avedon, D., J. R. Levy, *Electronic Imaging Systems*, McGraw-Hill, New York, 1994.

(9)   Yager, Tom, *The Multimedia Production Handbook for the PC, Macintosh, and Amiga*, Harcourt Brace, New York, 1993.

(10) Roca, R. T., "A Unified Network Management Architecture," AT&T Bell Labs, IEEE Network Operations and Management Symposium, Chicago, 1988.

(11) Freeman, M. B., "Providers Ensuring Network Performance," GTE Atlantic Operation, IEEE Network Operations and Management Symposium, Chicago, 1988.

(12) Woolston, D. S., "Architectural Support of Network Management: An Alternate View," Novell, IEEE Network Operations and Management Symposium, Chicago, 1988.

Chapter

7

Multimedia Technologies: Coding and Compression

Chapter Highlights:

The conversion of analog data into digital forms that are suitable for transmission involves a string of processes that deserve their own chapter for discussion. There are three main topics associated with this area of concern. The first is the data conversion process, which entails the creation of a raw digital signal from analog wave shapes. The next topic is that associated with the coding of information so it can be represented as part of a modulated transmission signal. The last major topic in this chapter is that of compression. Compression is a feature of the signal such that the information content is represented with as few bytes in data as possible. The coding discussion of this chapter is topical in that it covers several hot topics, among which are the JPEG and MPEG standards thought to provide the best algorithms for the efficient transmission of still images and video alike. Also discussed in this chapter is the next stage of coding, which is essentially a neural network-style method of coding and conveying imagery.

7.1 Data Conversion

The first, and probably most important, coding and compression issue to be addressed in the area of multimedia technologies is that of data conversion, because the conversion process is key to correct signal collection and representation.   Normally, data conversion is performed assuming a linear relationship between the input analog signal and the output digital representation of that signal. The resultant digital representation of the analog signal is referred to as pulse code modulation (PCM), because each sampling point or pulse results in a code generation activity.   The typical conversion paradigm is illustrated in Figure 7-1. The transformation process simply provides a translation between the incoming analog signals and their digitally coded equivalents.

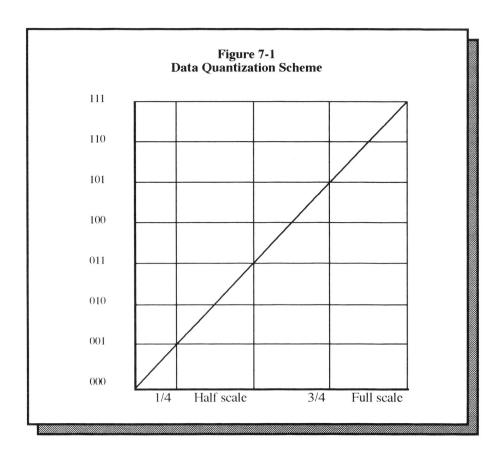

Figure 7-1
Data Quantization Scheme

The digital conversion process, illustrated in Figure 7-1, works as follows. The input signal is presented to a *sample and hold* digital conversion circuit. A sample and hold circuit is an electronic circuit that samples different points along a continuous analog wave shape and converts these sampled points to a digital representation whose size and granularity is determined by the number of bytes allocated to each sample. For a 1-byte converter, the maximum range that can be represented is 255. This means that the digital representation can parse the analog input into 255 parts. The conversion error, therefore, is no less than 0.0039, or $1/255$.

If the value of the input sample is measured and found to be one-fourth of the full-scale range possible for that analog input signal, then the digital equivalent of one-fourth is recorded by the circuit. This digital equivalent, for a 3-bit analog-to-digital (A-to-D) converter, as illustrated in Figure 7-1, will be 001. For this example, the entire analog input is divided into 8 segments, which can be represented by a 3-bit output signal (2^3). If the input scale were divided into 32 segments, the output code would have to be 5 digits (2^5) long.

The larger the code, the finer the granularity of the segmentation, and therefore the better the resolution of the process. The resolution error is that percentage of the total signal that could contribute to the ambiguity of the resultant sample. For example, if a signal sample arrived about halfway between one digital value and another, the possible error could be one part out of the total range of the conversion range. Thus, for a 3-bit conversion system, the possible error could be as much as $1/8$, or 12.5 percent. For a 5-bit converter, the possible error is $1/32$, or 3.125 percent, and so on.

For an analog value less than one-sixteenth full scale, the circuitry interprets this input and assigns a value of a digital 000. If the full-scale deflection is presented to the input, the digital value of 111 is assigned. One question always of concern in any discussion of A-to-D converters is that of resolution. For a 3-bit converter, the signal range is divided up into 7 segments, and for a full-scale deflection of 10 volts, each segment occupies a range of approximately 1.43 volts. At the point dividing this range into equal halves, that is, 0.714, the digital equivalent of this signal could be either of 2 digital values. For example, an input of 2.144 could generate an output of either 001 and 010. This means that the resolution of a 3-bit converter, for a 10-volt range, is no better than 1.43 volts.

7.2   Pulse Code Modulation (PCM) Techniques

Pulse code modulation is a data collection technique whereby the analog input signal is sampled, and a digital representation of that signal is generated which corresponds to its analog value. Most PCM technologies generate fixed-length codes between some arbitrary minimum, such as zero, and a maximum. The maximum swing, or range, of the sampling system is referred to as the *dynamic range* of that system. The resolution of the sampling device refers to the maximum number of bits allocated to each sample. Thus, an 8-bit sampling device can resolve any incoming signal to 256 different levels. A 12-bit sampler can resolve an incoming signal down to 4096 different values. A block diagram of the generic PCM method is shown in Figure 7-2.

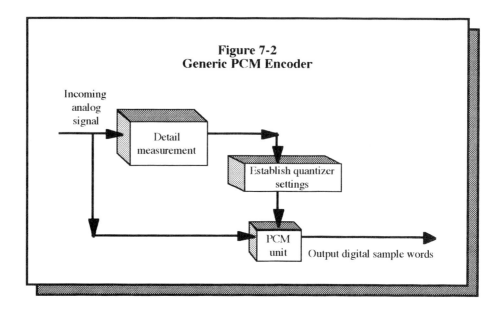

Figure 7-2
Generic PCM Encoder

Coding can be an effective compression tool if implemented so as to accommodate the bandwidth of the transmission medium and/or the imagery data rate(s). The incoming analog signal is analyzed for its detailed parameters. The results are used to set the PCM sample and hold circuit that generates the PCM values. These values are, in turn, used to format the data words output by the system. The sample and hold circuitry generates values that are relative to the dynamic range derived for the segment evaluated. Cited below are 2 of the more popular methods of utilizing PCM.

7.2.1 Adaptive Pulse Code Modulation (APCM)

The basic approach used in APCM is to segment the image, determine its maximum value for that segment, and measure the values of the pixels within that segment relative to the maximum previously found. The APCM encoder unit has the specific configuration as shown in Figure 7-3. Incoming video signals are segmented into small 3-by-3 or 4-by-4 video segments. For each frame matrix, the dynamic range is determined and the minimum value for this sample group is calculated. The minimum code length is derived and fed to the quantizer, while the minimum value is subtracted from the incoming sample. The quantizer performs a PCM conversion on the analog samples, and passes these digital samples to an interface for further processing.

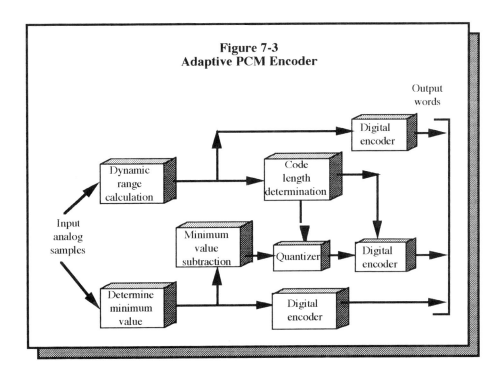

Figure 7-3
Adaptive PCM Encoder

7.2.2   Differential Pulse Code Modulation (DPCM)

Another primary usage for PCM techniques is to use it for the differential coding of the image signal. A DPCM encoder operates as follows. The video frame is broken into image samples of some number of lines and pixel widths. A prediction is then made of the same image sample space in the next frame. This can be expressed as:

$$\delta[m,n] \quad = \quad p[m,n] - \varepsilon[m,n]$$

where

$\varepsilon[m,n]$	=	the prediction matrix for the image sample
$p[m,n]$	=	the original video frame image sample
$\delta[m,n]$	=	the prediction or error in interframe movement

At the receiving end, the decoder adds the prediction to the error transmitted from the encoder, and the following function is obtained:

$$\pi[m,n] \quad = \quad \varepsilon[m,n] + \delta[m,n]$$

where
$$\pi[m,n] \quad = \quad \text{the reconstructed image sample which is intended to be a replication of the original sample}$$

If $\pi[m,n]$ and $p[m,n]$ correspond exactly, then the prediction was entirely correct. The method of arriving at the prediction is still to be determined.

7.3   Information Coding

7.3.1   Loss versus Lossless Algorithms

Data encoding and decoding schemes, for both coding and compression requirements, have been developed to handle the data that they represent. There are 2 basic types of data encoding/decoding approaches, one being what is referred to as *lossless* algorithms, and the other is that which allows loss in the encoding process in favor of a bandwidth advantage. The *lossless* algorithms are those that purport to convey all of their information without that information suffering loss of content. This faithful reproduction of the decoded information is accomplished through bit rate reduction, such that redundant elements of the signal are eliminated. Additionally, in some cases the signal received at the terminating end need not be entirely recovered in order for that signal to be reproduced.

The relationship between the bit rate and the encoding problem can be explained with the following example. If a digital signal, such as a monochromatic TV image, needs to be transmitted from point A to point B, the signal could be reduced in bit rate by a technique that calls for the first pixel to be transmitted as its actual value and trailing pixels coded so as to reflect only differences in their values between this starting value and succeeding values. The range of digital imagery pixels is nominally 0 to 255, or 2^8. Each pixel, therefore, is represented by one 8-bit byte. As the stream of data is processed 1 pixel at a time, each pixel, after the first one which is used as a reference value, is assigned a value based upon its relationship to the reference value. This technique is a version of what is referred to as *differential pulse code modulation (DPCM)*. It is explained in diagrammatic form in Figure 7-4.

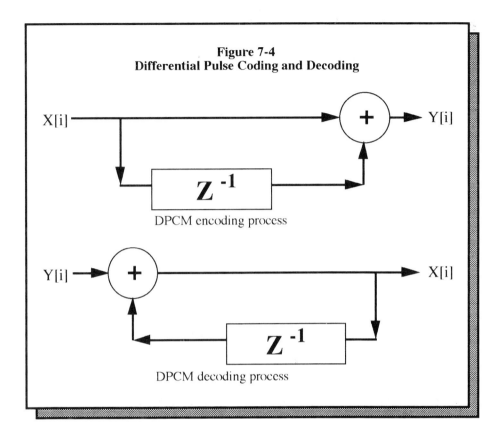

Figure 7-4
Differential Pulse Coding and Decoding

DPCM encoding process

DPCM decoding process

Figure 7-4 shows both the encoder and decoder for a DPCM method of coding. In the encoding process, the original signal source comes into the encoder as X[i]. The first signal element, at the first slot or time interval, passes through the coder without modification, because Z^{-1} is a delay element in the filter. The next element, however, is added to the first signal element to yield its resultant value, and so on. The process is reversed in the decoder, where the first element coming into the filter passes through without modification. Succeeding elements are modified based upon the value of the first element to yield the original signal acted upon at the encoder.

Algorithms that introduce loss into their compression and coding processes are those that reduce the signal beyond the point of minimum bit rate reduction, or the elimination of redundant signal elements. An example of this problem is when a signal that is normally represented by a 4-bit half-byte is compressed to a 3-bit representation. Clearly, the accuracy of the original signal is compromised by this reduction, which could be viewed as either a bit rate reduction or an elimination of redundancy.

7.3.2   Interframe Coding

Motion picture source compression relies upon the assumed attribute of temporal redundancy to encode the digitized signal. This reliance works only when 2 or more pictures are subjected to the encoding algorithm. This process is referred as *interframe coding*, because each video picture is known as a frame, as opposed to individual picture coding, which is known as *intraframe coding*. The basic approach to such encoding methods is described as follows, and is keyed to Figure 7-5.

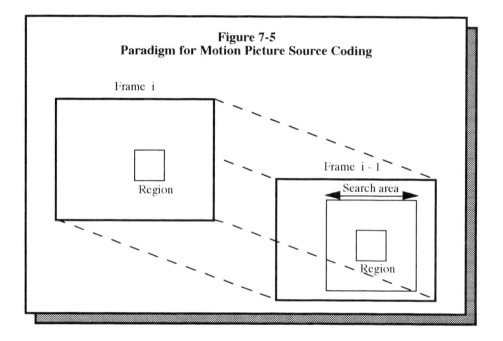

Figure 7-5
Paradigm for Motion Picture Source Coding

Frame i

Region

Frame i - 1

Search area

Region

The processing first begins with the frame being divided into a large number of small regions. Divisions of 8 by 8 or 16 by 16 pixels each are often used for analysis. A search area in the first frame is selected containing the first region of interest. In frame i - 1, a search area is defined and a region within that is selected. The next frame, frame i, is now scanned to determine the best location fit for the region of interest within the new frame space. Clearly, the decoding process must be able to recognize this shift, so the decoder is sent such ancillary information in the compressed data stream. This motion shift between frames i - 1 to i is referred to as the *motion vector*. The process of determining this vector is known as *motion estimation*, and the prediction error is referred to as the *displaced-frame difference*.

There are some risks associated with this and similar approaches to motion estimation, among which are that the image elements cannot move significantly between frames, otherwise the system falls apart, and that the search area must be kept sufficiently large in order to allow for significant movements in the spatial domain. If the search area is defined to be the entire frame space, the process is slowed down, and a slow-reacting algorithm will impede the rate of image transfer. If, on the other hand, the search area is defined to be the size of a region, or slightly larger, a fast-moving image will not be encoded properly, and much of the image motion will be lost. Faulty optimization would make a movie, such as *Spartacus* or *Star Wars*, impossible to watch, because most of the action would be garbled, as most of the scenes in these movies are loaded with action. Unfortunately, an optimized algorithm cannot overcome the data rate limitations of the transmission system, such as copper wire. These issues will be explored more fully later.

The process analyzes and determines differences between object movements between frames. This effect is illustrated in Figure 7-6. The 2 frames have overlapping objects that appear to have moved between frames. The actual movement is reflected in diminished object representation as indicated on the right side of Figure 7-6. The larger the movement, the more error there is to deal with, whereas the more static the image, the less error there is to process. Thus, action-packed performances have more displacement, i.e., error, to deal with that consumes more bandwidth than, say, talking heads.

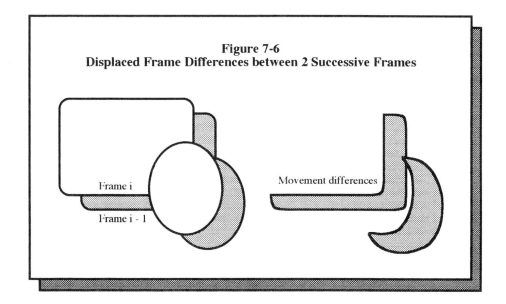

Figure 7-6
Displaced Frame Differences between 2 Successive Frames

Frame i

Frame i - 1

Movement differences

The encoding process scans the search area to locate the most similar set of pixels to those originally chosen for analysis. The derived location for the set of pixels under investigation, in terms of direction and distance from the previous pixel array, determines what is called a *motion vector*. Additionally, the degree to which the region is faithfully reproduced in the new location, is established by such measures as the *mean squared error* (MSE). The original region is compared to the succeeding region to derive the error between the two. This effort to define the most similar region, using the inspection area for investigation, is known as *motion estimation*. The mean squared error is defined as:

$$MSE \quad = \quad \varepsilon \, \{(X[i] - x[i])^2\}$$

where

$$\varepsilon \qquad \qquad = \qquad \text{statistical expectation}$$

$X[i], x[i] \quad = \quad$ output and input sequences of pixels respectively, where i = the identification of each pixel

The prediction error, referred to as the displaced-frame distance (DFD), is quantified and encoded for transmission. The motion compensation algorithm used for prediction purposes is called the differential pulse code modulation (DPCM). It is based upon the Z transform, which is essentially one time interval delay between incoming values. The basic approach is illustrated in Figure 7-7.

As the incoming signal traverses from left to right, a copy of it is delayed one time interval such that, after one delay, the copy is added to the pass-through version of the signal. At the receiving end, the signal is again shifted by one time slot, such that the delayed component is canceled, and the original signal content is recovered.

7.3.3   Motion Picture Experts Group Coding

A variation on the interframe coding theme is that coding scheme called MPEG. For several years, the Motion Picture Experts Group (MPEG) has been studying and developing various motion-compensation coding algorithms that are based upon interframe, time displacement coding techniques. MPEG is a subset of the International Standards Organization (ISO). The furor over HDTV, fiber optics, increasing data requirements over existing copper plant, and anticipated market needs for more information and entertainment sources has sparked increasing emphasis upon compression algorithms and codecs (coder/decoders). The emphasis has been on compression to achieve reduced transmission rates of between 1.2 Mb/s and 10 Mb/s. The basic approach taken to motion compensation prediction is illustrated in Figure 7-8.

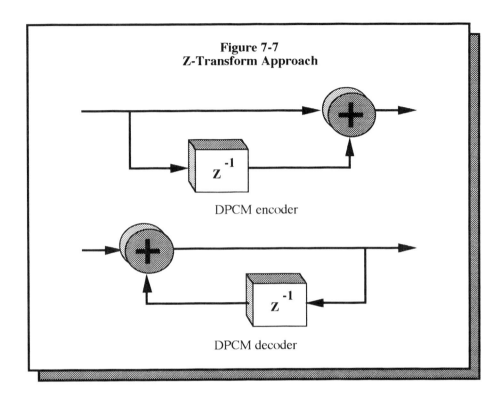

Figure 7-7
Z-Transform Approach

DPCM encoder

DPCM decoder

7.3.4   Linear Predictive Coding (LPC)

Signal encoding and decoding is a technology employed to generate bit streams suitable for transmission over the communications system of choice, typically copper, coax, microwave, etc. After data is passed through an A/D converter, it is subjected to a specialized prediction algorithm. A prediction of the next value to be received is based upon the previous value, or upon a group of previous values.

If the pixel value predicted is correct, the circuitry returns a zero that is output from the comparator. Thus, the actual value of the pixel need not be transmitted, but rather an indication of its relationship with previous values collected. If the prediction is incorrect, a one is returned, and the new pixel value is transmitted as a new reference point. A schematic diagram of the concept behind the linear predictive coder used in video transmission is shown in Figure 7-9.

The predictor is a device that can escape normal logic. Its determination of what the next symbol value will be could be based upon its previous run of successes, or possibly what the run length was in the previous horizontal line. Whatever its method, a random call by its algorithm would expect to generate success about half the time.

Linear predictive coding and decoding systems have success rates of between 10:1 and 40:1 in data reduction. For example, a digital voice signal that has been processed through a 12-bit A/D converter with a sample rate of 8000 samples per second will generate a data rate of about 100,000 bits per second. LPC will reduce this requirement to 2400 bits per second for intelligible speech, and about 9600 bits per second for high-quality speech.

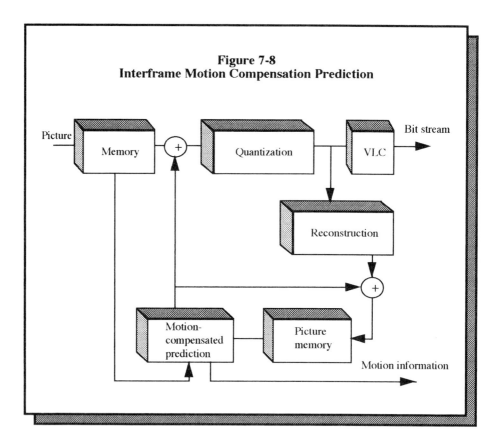

Figure 7-8
Interframe Motion Compensation Prediction

The essential ingredient of the encoding/decoding system is its predictive component. The basic theory associated with this device is that, no matter what its method of prediction for encoding, it must reverse that method for the decoding process. A random prediction algorithm would, over the long run, predict correctly half the time. The prediction process is one that proceeds as follows:

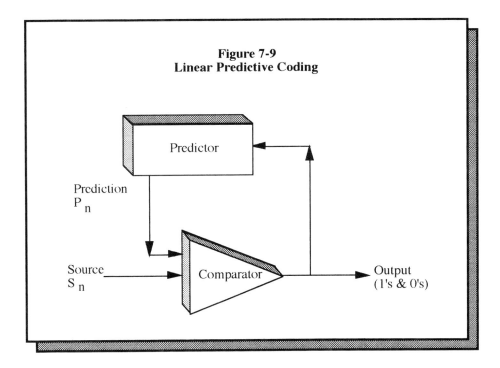

Figure 7-9
Linear Predictive Coding

(1)   The input data stream is examined one data sample at a time.

(2)   The predictor makes a calculation as to whether the next input
       data sample will be the same as or different from the previous
       value examined, or from a grouping of previous values
       encountered.

(3)   Correct predictions will yield a zero output from the LPC circuit,
       and incorrect predictions will yield up a value of one.

(4)   The compressed output is a message that provides the value of the
       run and the number of pixels assigned to that value.

Thus, for a display that has 8-bit gray-scale color resolution, a run of 255 pixels,
all having the same value, would require two, 8-bit words, whereas an
uncompressed stream of pixel information would require 255 8-bit words. This
equates to a compression ratio of 127:1, or $255/2$.

Figure 7-10
Comparison of Uncompressed and Compressed Video Frames

Uncompressed Video Frame #1

Compressed Video Frame #1

Figure 7-10 (cont'd)
Comparison of Uncompressed and Compressed Video Frames

Uncompressed Video Frame #2

Compressed Video Frame #2

Figure 7-10 (cont'd)
Comparison of Uncompressed and Compressed Video Frames

Uncompressed Video Frame #3

Compressed Video Frame #3

Figure 7-10 (cont'd)
Comparison of Uncompressed and Compressed Video Frames

Uncompressed Video Frame #4

Compressed Video Frame #4

7.3.5   Run-Length Coding

Many of the algorithms used for video compression are based solely upon a running count of like-valued pixels in the data stream. Runs are identified by the value of the run and its duration. Such approaches are generally lossless as opposed to lossy, where the quality of the playback is sacrificed. Lossy video results because the decompressed pixel stream does not faithfully reproduce the original input stream. Figure 7-10 shows a series of frames from the same video sequence. Depending upon the rate of action within the frame, the frame rate of the compressed imagery may have to be reduced in order to keep up with the incoming frame requirements. The next section discusses these issues more fully.

The first, third, fifth, and seventh frames are uncompressed frames of a video stream that were captured and digitized at 30 frames per second. The even set of frames have been compressed using a run-length algorithm. The reader will observe that each frame scene shows drastically different treatment based upon whether the image was subjected to some form of compression. The reconstruction of compressed images tends to show certain distortions accompanied by the appearance of a washed-out image with resulting poor image details. Sometimes certain other grainy features appear that further distort the image reconstructed, such as streaking, where narrow white lines or streaks appear accross the face of the image.

There are many other factors that affect the quality of the images reconstructed from compressed states in addition to the losses imparted to the scene by the compression algorithms themselves. These include the frame resolution, and frame size requirements, the frame rate at which the images were captured, and the speed of the processor, which may impose algorithmic restrictions upon the calculations and manipulations of the processor itself. Each of these is elaborated upon below.

(1)   The frame resolution, and size of the frame, dictate the detail which is to be reproduced. If the size of the reproduced frame is much larger than the pixel resolution provided in the original imagery, the resolution of the reproduced image will be relatively poor and detail within the image will suffer.

(2)   The frame rate of the video capture system may also dictate resolution by either preserving it if the frame rate is high enough, or degrading it if the frame rate is slow. Frame rates may contribute to the smearing or degradation of imagery as the information rate proceeds through the scenery.

(3)   The processor instruction execution rate may determine the types of processing possible on an image of specific dimensions. For instance, complex image processing algorithms may consume a considerable number of instructions per second while simple filtering algorithms may be within the capacity of the processor's speed.

7.3.6   Huffman Coding

This approach to the coding of signal samples is a variable-length coding system that is based upon the frequency distribution of expected samples. For example, if one were to establish a coding system for the American version of the English alphabet, it would first be necessary to generate a frequency table for a sampling of American English words. Such a table would result in the following frequency percentages using some sample of the language, such as Lincoln's Gettysburg Address:

Letter	Frequency (percentage)
E	13
T	9
N/R/O/A/I	7
S	6
D	4
L	3
H/C/F/P/U/M	2
Y/G/W/V/B	<2
X/Q/K/J/Z	<1

This method of character frequency analysis is used in cryptanalysis to break ciphers used in secure communications. It is also useful for, and has a place in, the area of data compression as well. Since the letter E occurs most often in the English language, its transmission takes up most of the available bandwidth, to the extent that its occurs most often in the words conveyed, therefore most of the characters in the data stream tend to be the letter E. If it were possible to transmit this, or any, symbol in a way that utilized fewer bits, then the symbol would occupy less of the bandwidth available for transmission. This can be done by essentially establishing a code that is the inverse of the symbol's frequency. In other words, the higher the frequency of utilization, the fewer the number of bits used to represent the character of interest.

Thus, the letter E, which occurs most often, would be coded with the fewest bits; while the least occurring letters, i.e., X, Q, K, J, and Z, would require the most bits to convey. This inverse relationship is based upon the fact that the most frequency values are transmitted most often, and, therefore, bandwidth can be conserved by using fewer bits to code their existence, while less frequent values can be relegated to more bits because they occur much less often.

The same logic can be applied to pixel values that are transmitted to a monitor for display. Extreme values of color would be represented by larger numbers of bits, while basic colors or shades of gray would occupy fewer bits per pixel value. As an example, suppose we have a modest intensity range for possible pixel values of, say, 16 levels, or 2^4. A scenario for these 16 level values, obtained from a random number table, could result in the following tabulated values:

Value	Frequency	Value	Frequency
1	8	9	5
2	4	10	7
3	13	11	10
4	2	12	14
5	4	13	3
6	6	14	4
7	9	15	5
8	4	16	2

Value #12 is the most frequently used, and therefore would have the least number of bits to represent its existence, whereas values #4 and #16 are the least used and therefore would require the most number of bits. The following table could represent these adjusted value structures:

Value #	Word Structure
3/11/12	01/10/11
1/6/7/10	001/100/101/111
2/5/8/9/14/15	0001/0110/0111/1000/1100/1111
13	0111
4/16	0011/1010

The resulting coding system would compress the nominal data rate for this pixel distribution to about 74 percent of the uncompressed version.

7.4   Complexities of Information Coding Algorithms

7.4.1   Delays

Coding algorithms and devices often introduce a coding delay into their respective systems. This delay is a function of the number of pixel values to be processed within the coding process and the algorithmic steps through which the coding algorithm proceeds. For instance, if the number of pixels to be coded is 20 and the code steps through 100 instructions at about 5 clock cycles per instruction, or operations per unit of time, and each pixel is clocked at 1 clock cycle per pixel, then the coding delay will be at least 520 clock cycles. Obviously, the actual application determines the extent of the delay.

7.4.2   Complexity

The algorithm that executes the coding scheme, while a function of the number of operations per second, has a complexity associated with it so that the complexity determines its average operational duration. There are trade-offs between the operational rates and the quality of the coding that can be achieved. Thus, the basic trade-off is one of lower complexity at the expense of either lower quality or higher bit rate. Commercial coding systems can achieve more complex coding schemes than consumer devices, but the basic idea is still the same, i.e., to provide efficient encoding for information transfer.

7.4.3   Sensitivities to Algorithms and Channel Errors

The source coding algorithms rely upon stable data streams in order to achieve consistent results, also referred to as reliable output. Coding algorithms are developed based upon many assumptions about the data to be processed. As this data varies, the efficiency of the coding schemes varies as well. Some coding algorithms are more effective than others in resisting such variations. These abilities of the coding schemes to handle such conditions determine their relative attractiveness for source coding.

Another issue associated with the problems of sensitivities is that relating to channel errors. As coded data is transferred between stations, channel errors may not be corrected. As a result, erroneous information packets or data streams are received and additional distortion is introduced into the data stream as a result of shifted and/or erred data packets. Some coding schemes are more sensitive to error conditions than others. For example, decreased bit rates will usually increase the sensitivity of the scheme to undetected errors. In some cases, however, if channel errors are detected, it may be possible to conceal or correct such errors after source data decoding.

Technical Note 7-1
Data Coding Versus Data Compression

Data encoding has a history as long as telecommunications itself. The ancient Romans used signal fires, in a fashion similar to that exercised by the Native Americans. The Romans used smoke signals to relay messages from one end of the Roman Empire to another in a matter of hours. They used a coded signal system to convey information about political and military matters from Rome to its outlying provinces. Later the English used a system of semaphores to relay messages across England. More recently, Morse used long and short timed closures of an electrical key device to convey specific letters of the alphabet over transmission lines that were analogous to normal telephone lines. All of these examples of signaling represent data encoding, that is, a representation of the intended message by some substitute form [3].

Compression is a relatively recent innovation that is intended to carry the message by formulating it in a fewer number of symbols than originally assigned in the coding schema. There have been many techniques developed to perform such tasks, and some of these will be summarized here. The reasons for pursuing such techniques are, effectively, to multiply the bandwidth ability of the transmission system to carry the original message symbols, and to conserve the storage media used hold the data of interest. Data compression, among other things, lowers transmission costs because less time is required per transmission per unit message, assures a certain amount of privacy to prevent disclosure from casual observation, and reduces the storage space requirements for compressed versus uncompressed images.

7.5   Data Compression

The term *data compression* will be used here to refer to all the media, i.e., text, audio, and image compression. We must, however, not confuse data compression with data encoding. This distinction is discussed in Technical Note 7-1.

In a technical sense, the concept of data compression is embodied in the idea of the effective information transfer ratio (EITR). This is the ratio of the actual data transfer to the nominal communications line transfer rate, or symbol transfer rate/line transfer rate (in symbols per unit time).

$$\text{EITR} \quad = \quad \frac{\text{real data transfer rate}}{\text{nominal data transfer rate}}$$

This value is similar to another term, used frequently, known as the *compression ratio*, which is the ratio of the length of the original data string to the compressed data string.

$$\textbf{Compression ratio} \quad = \quad \frac{L \text{ original data string}}{L \text{ compressed data string}}$$

Another term used with regard to data compression is the *figure of merit*, which is the reciprocal of the compression ratio, and is defined as the length of the compressed data string divided by the length of the original data string.

$$\textbf{Figure of merit} \quad = \quad \frac{1}{\textbf{Compression ratio}}$$

There are 2 principal means of data compression, one being physical and the other logical. Physical data compression is the process of eliminating as much of the data as possible prior to entering the transmission system. Logical compression is the replacement of alphanumeric categories with numeric or binary substitution equivalents.

There are a number of commercially available compression software packages on the market today, as well as proprietary algorithms used for such products as ADSL and HDSL. Specific data compression techniques that have been developed and are being used are summarized as follows.

7.5.1   Null Compression

Null compression is one of the earliest instances of a data compression implementation, and is even used today by IBM and IBM emulators in the BISYNC protocol. It functions by detecting strings of blanks, or nulls, in a data stream and then replacing them with a special format that signals the null count using a special control character followed by the actual number of nulls in that string. For example, the ASCII data string, A B C D # # # # L M N O # # # # # X Y Z, would be reduced to the null compression form, A B C D S_c 4 L M N O S_c 5 X Y Z. The result saves 5 spaces overall between the 2 strings of nulls as depicted by the # signs. S_c is the special control character used that signals to the software that the null count follows, followed by the numeric value of the number of nulls in the run.

7.5.2   Bit Mapping

Bit mapping is a technique that, for every 8 characters, generates a special character to identify null character positions and deletes these characters. The special character is generated by indicating in its bit stream where the nulls were positioned. It is then inserted into the data stream during the encoding process.

It is later used at the reception end to replicate the original stream of characters during the decoding process. Thus, a string that has the sequence

$Data_1$ - $Data_2$ - Null - Null - Null - $Data_3$ - Null - $Data_4$,

in the bit mapping technique, will yield a sequence of 8-bit characters that carry the non-null values plus the special character that contains the locations of the null characters. Thus, the sequence of 8-bit characters will look like this:

$Bitmap$ - $Data1$ - $Data2$ - $Data3$ - $Data4$,

where
        Bitmap = 11000101

Therefore, the bit map compression ratio can yield as much as an 8:1 compression ratio. On the average, however, one can expect a compression ratio of 8:5, because half of the string may be null characters, and half are non-null characters. Therefore, the non-null characters are retained, and the remaining character identifies the placing of the null characters.

7.5.3   Run Length

The run-length technique is particularly suited for image compression, because images tend to have long runs of pixels with the same color or gray-scale values. A *pixel* stands for picture element and is a point or dot on the display screen upon which a color or gray-scale value is exposed.

If the image is a still picture, the persistence of the image is sustained by the refresh rate of the video electron gun. If the image is one of a motion video stream of frames, the image of one frame lasts only until its next frame is to be displayed.

An image compression standard known as JPEG was devised to compress still images. Using a frame grabber, and software that can compress and decompress such acquired video frames, Figure 7-11 illustrates the differences between an uncompressed video frame image and a JPEG-compressed image, from the same tape, that has been subsequently decompressed, focusing on the same scenery for this comparison. The result is a noisier image for the JPEG image recovery. Notice, for example, that the airplane tire in the selected image has different numbers and larger white marks on the JPEG-reconstructed frame than the frame that was not compressed for storage.

Figure 7-11
Comparison of Uncompressed and JPEG Compressed Image

Uncompressed Video Frame

JPEG Compressed Video Frame

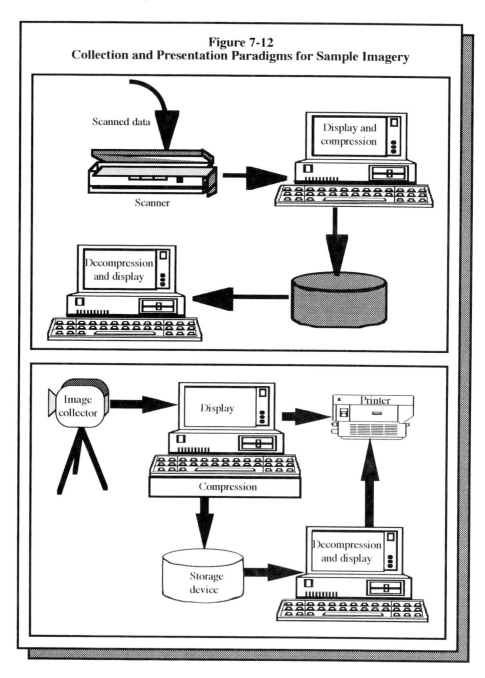

Figure 7-12
Collection and Presentation Paradigms for Sample Imagery

The reason for the less-than-faithful reproduction of the decompressed image as shown in Figure 7-11 is that there is a certain amount of smoothing in order to compress the image as much as possible. The algorithm has a threshold for compression of a running set of pixels such that if the algorithm finds a number of pixels that fall below the threshold value, the run-length compression operation will be overlooked and the pixels will be incorporated into the next run-length calculation. Another indication of the JPEG compression algorithm imperfection can be seen in the detail shown of the fuselage between the propeller and cockpit. The JPEG-compressed image appears to be more washed out than the uncompressed image.

The comparison shown in Figure 7-11 and those to follow in succeeding chapters are all based upon the collection, compression, storage, and retrieval paradigm illustrated in Figure 7-12. The method of presentation used here is, first, to collect an image and display it in uncompressed format. This is illustrated in the first frame of Figure 7-11. Second, that same image is stored in some type of compressed format, and then retrieved, decompressed, and displayed. This process is illustrated in the second frame of Figure 7-11. The second half of Figure 7-12 shows the method of collection and retrieval used here.

7.5.4   Diatomic Encoding

Diatomic encoding is a process of substituting a special byte sequence in place of 2 bytes that occur frequently enough to warrant a special byte to represent these 2 paired bytes [3]. Thus, the maximum compression ratio possible for such an encoding technique is 2:1 or a 50 percent data reduction. The general formula for diatomic encoding is:

$$\text{Char}_1 \ \& \ \text{Char}_2 \ -> \ \text{Char}_{spec.}$$

Diatomic data compression, along with the other algorithms discussed here, may also be mechanized in a hardware implementation as well as by software means. A hardware implementation could be explained by use of an example, such as that offered by a time-division multiplexer. Such a multiplexer takes low-speed sources and combines them into a high-speed aggregate by assigning each input to a particular time slot on the output side. Any compression at the input side results in no effect on the high-speed side, because a specified time slot and bandwidth is always allocated to that input whether or not it is subjected to compression algorithms. On the other hand, a different type of multiplexer, known as a statistical multiplexer, reacts differently. A statistical multiplexer only takes data that is actually presented at the inputs, and multiplexes it with other incoming data in time-slotted fashion and allocates the bandwidth needed. Therefore, a compression of one or more inputs will have a direct effect upon the capacity of the output side and its ability to carry more input channels than a conventional multiplexer.

This type of compression communications was originally implemented to accommodate data traffic. Some vendors compress the high-speed side, while others compress the incoming channels. The most common and easiest compression method to employ is the diatomic encoding technique.

7.5.5   Pattern Substitution

The pattern substitution approach to compression is similar to methods used in cryptology. One of the methods used in cryptology is called code cryptology, where substitute words, word strings, or nonsense character sets are substituted for words or word groups in the original message. This type of technique is sometimes also applied to data compression, where commonly used character sets, or multiple byte pattern sets, are substituted for special bytes that represent the substituted strings. For example, the phrase

"Now is the time for all good men to come to the aid of their country" might be coded as

"xm yt ab zp lw ms ne rf bq aj bq da cx zu uy ge"

In order for pattern substitution to work properly, a dictionary must be created that catalogs the bit patterns to be substituted. Each bit pattern of interest is then associated with a substitute bit pattern representing the original pattern. This cross-reference specification can then be programmed into the relevant software module that performs the encoding task. At the receiving end, the same table is embedded in the receiving module that converts the substitution back to its original, meaningful form.

7.5.6   Relative Encoding

Relative encoding is a method used in telemetry systems. It became popular at the outset of the space age, when satellite and missile information had to be relayed back to the ground for evaluation and vehicle commanding. This form of encoding is used quite often in telemetry coding. The theory of operation for relative encoding is that the original signal will remain relatively constant or deviate only within a narrow range during the period of sampling. Should the signal begin to vary outside this overall range, a new base reference point is established and the subsequent reports continue to report only deviations. An example is illustrated below.

Original signal string
100   100.1 99.8 99.7   99.9   100.1 100   100.2 100.0 100.3 100.1 99.9

Relative encoded string
100 +0.1 -0.2 -0.3 -0.1 +0.1 0   +0.2 0   +0.3 +0.1 -0.1

Notice that all values are offsets or deviations from the previous value. Coding requires much fewer bits to reflect these deviations than the entire value indicated in the original signal string. A half-byte coding scheme could accommodate an error range of 1.5 using increments of 0.1 above and below the reference level.

7.5.7   Digital Facsimile

Digital facsimile has become a mainstay of business since 1986 when the Group 4 Standard was established and FAX machine prices started to drop rapidly. One of the principal advantages of FAX, from a bandwidth point of view, is that the image consists of either white or black spots on the hardcopy surface. The other major advantage of FAX is that it can be transmitted at variable rates without destroying the content of the message, unlike video, which depends upon a constant high-speed rate in order to adequately convey its impression. FAX can be transmitted slowly or at full line rate even though that rate may take some time, and the nature of the content will not be harmed in the least.

A capability that supports the conveyance of text, graphics, and imagery to a viewer at some remote location, i.e., outside the sender's local area network, for example, is key to the furtherance of any customer's multimedia awareness, and its potential. The incorporation of speech and video annotation is also possible with the new compression techniques for digital audio, and at least slow-scan TV, as described in the previous section.

The basic concept behind digital facsimile, or FAX, is the scanning of an image on a line-by-line basis. The scanned image, after the initial hand-shaking interchange, is continuously output to the receiving FAX during the scan process. The key issue involved in FAX is resolution. Part of the resolution obtainable by the FAX system is determined by the granularity of each scan, but with a narrower scan comes the requirement for more lines to be transmitted to the receiving FAX unit along with the bandwidth needed. The other major component of the FAX resolution problem is that of scan points per line, or picture elements (pixels). In many respects, FAX is similar to the operation of TV. The NTSC standard calls for 525 lines per image reproduction frame. The FAX standard calls for 100 scan lines per inch in order to reproduce normal typewritten text. A normal scanned page of 8 $1/2$ by 11 inches requires 850 scan lines, and each scanned line contains 1730 pixels. The data required for each page is therefore 1,470,500 bits.

Without compression, a page of text would need over 300 seconds to transmit at 4800 bps, and about half that at 9600 bps. Because most of the page is not used for conveying the image components, compression techniques, such as run-length coding, could create a compression ratio of 20:1 or better.

The run-length technique encodes the existence of runs of white background or black spots to be transmitted from the originator to the receiver. Another technique commonly used is to transmit only the differences between each current scan line and the previous one. This technique also provides a 20:1 or better compression advantage. Another method used is called the Huffman coding technique. This is a statistical coding method that examines the frequency of the various bit patterns transmitted and assigns substitute codes for such data sets.

7.6   Digital Audio Compression

The world of audio compression and digital manipulation and conversion of analog signals has reached new heights in recent years. Starting with the use of CD-ROMs to reproduce and capture the quality and fidelity of musical audio renderings with their high-frequency components, the popularity of digital music has soared over the past 10 years. The system aspects of such technological innovations can be reviewed briefly here to acquaint the reader with the basic elements of this side of the multimedia discussion.

Audio processing is divided into 2 major categories, one being the compression contribution, and the other being the digitization contribution of the processing problem. Compression can range from none at all to a 3:1 or 6:1 ratio. The types of techniques used vary and are discussed elsewhere in this book in some detail, but they include run length, null suppression, and relative encoding. The digitalization segment of audio processing may utilize either an 8- or 16-bit word. The size of the digital transformation word determines the granularity, or accuracy, of the conversion. It is chosen by virtue of the type and quality of the A/D hardware used.

The analog input audio signal may be passed through a bandpass filter of from 150 Hz to 48 kHz depending upon the quality of the reproduction required. The reader can see that, as the dynamic range of the recording requirement increases, the 8-bit word used to store this range becomes more and more coarse, i.e., it must store a larger range of values. Additionally, a 16-bit word, while also accommodating a larger range of values, will yield a much smaller absolute value range.

For example, the hertz per digitization level for an 8-bit word is:

$$\frac{16\,\text{kHz}}{256\ \text{levels per 8-bit word}} \quad = \quad 62.5\ \text{Hz per 8-bit word}$$

For an equivalent 16-bit word, the digitization level is:

$$\frac{16 \text{ kHz}}{65536 \text{ levels per 16-bit word}} \quad = \quad 0.244 \text{ Hz per 16-bit word}$$

Clearly, the integrity of the signal is more nearly maintained with a 16-bit digitization level than an 8-bit level. Another parameter is the sampling rate of the signal. This rate is determined by the frequency components in the analog signal. According to the Nyquist criterion, the sampling rate for an analog signal is 2.2 times the highest frequency component of the signal. Thus, for a 16-kHz dynamic range, meaning the frequency range over which the signal oscillates, and if the highest frequency is about this value, the sampling rate will be

$$2.2 \text{ samples/Hz x } 16,000 \text{ Hz/second} = 35,200 \text{ samples/second}$$

The accuracy of the samples is then determined by whether an 8-bit byte is used to resolve the analog signal, or a double byte is used.

7.7   Frame Compression Techniques

Compression techniques and algorithms react in ways not always predictable, because of their interactions with such parameters as frame sizes, frame rates, transmission rates, storage systems, etc. Some of the generic types of algorithms will be covered shortly, but the basic parameters should first be recognized. Nominally, the collection rate for video is 30 frames per second, using the NTSC standard that equates to approximately 480 by 260 pixels per display frame, and at least 255 gray-scale or color values. Various compression algorithms have been developed that attempt to compress video frames as well as still imagery.

7.7.1   Interframe Compression

There is an old expression to the effect that you cannot get blood out of a turnip. The existing copper outside plant available to the telephone companies all over the U.S. has prompted initiatives to develop compression techniques to accommodate video transmission using all of this copper. The motivations for this thrust have obviously been as much political as they have been technological. One of the principal groups at the forefront of attacking this problem has been the Motion Picture Experts Group (MPEG). The MPEG has developed various algorithms that have been labeled MPEG1, MPEG2, etc. These approaches rely upon the interframe method as discussed in previous sections.

At 30 frames per second, talking heads or upper torso shots generate little interframe displacement of pixels, or, said another way, pixel changes between frames. On the other hand, sequences with rapid movements, such as fighting or other high-speed activities generate large pixel displacements between frames, thus causing considerable interframe pixel changes to be transmitted. For limited bandwidth transmission systems, such as those that rely upon copper technologies, the problem now is the amount of interframe displacement that can be tolerated and still pass the image without loss. The trade-offs that must be faced are either to find some new technology that expands the bandwidth, to accept lossy imagery to be passed, or to pass fewer than 30 frames per second. If the frame transmission rate is decreased from 30 to something less, such as 15, the picture may not be degraded on a frame-by-frame basis, however, the video sequence will appear jerky and the scene will jump around in a very distracting way. Even slight reductions in rate will yield degraded imagery from time to time during the performance.

The MPEG standards accept lossy imagery for high interframe differences, i.e., rapid activity, through a combination of algorithmic compromises. As the video action increases, distortions become evident, and at some level of increasing activity within the scene, frames are skipped such that jerky movements result. Also, the sound accompanying the scene is coupled with the imagery. Loud sounds appear to create more of an impact on the compression process than softer ones. They are so tightly coupled with the available bandwidth that, in some cases, loudness appears to cause the picture to break up.

7.7.2   Intraframe Compression

Many techniques have been developed to compress individual images. The Joint Picture Experts Group (JPEG) has developed and published a compression standard, also known as JPEG, to compress such still frame imagery. Numerous other techniques have been developed over the years to process still pictures. These techniques have yielded highly efficient compressed frames. On the other hand, intraframe compression has the disadvantage of not being able to rely on previous frames for coding references.

7.7.3   Companding (Compression and Expansion of Data)

Companding is a compression technique that uses a logrithmic function to perform the compression and decompression process. Compander is a contraction for compressor/expander. It utilizes the usual steps of signal conditioning and conversion that include this logrithmic processing. It can be used to further reduce a compression process, such as run-length coding, so as to further reduce the signal of interest to about 67 percent of the original coded string. The algorithm is somewhat complex, but generally consists of the following procedure. For each 12-bit code word developed:

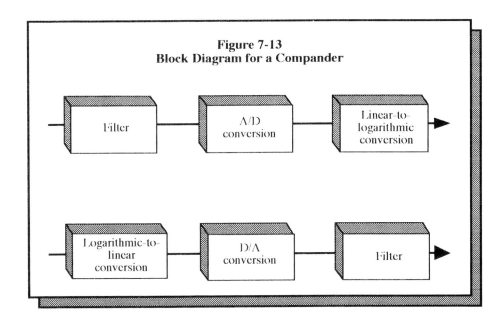

Figure 7-13
Block Diagram for a Compander

(1)   Add 33_{10} to the value of each code word.

(2)   Determine the leftmost bit that contains a one.

(3)   Subtract its position number, e.g., position 7, 8, etc., from 5, and record its value as binary number.

(4)   Take the 4 bits to the right of the bit position in which the leftmost bit was first located, and add them to the 3-bit binary number determined in (3) above.

(5)   If the number is positive, make the leftmost bit position of the 8-bit compressed word a zero, or a one if it is a negative number.

This type of system may be seen illustrated in Figure 7-13. In the upper portion of the figure, the incoming signal is conditioned by first filtering it to remove extraneous noise, then subjecting it to an analog-to-digital converter to reduce the signal to a digital form, and then applying the digital signal to the algorithm described above using the logarithmic function as the coding format. In the lower portion of the figure, the received signal is reversed by first reversing the logarithmic conversion format, changing the digital form of the signal to an analog form, and finally, filtering the signal to assure that the analog form is as clean as possible.

Summary

This chapter discussed the various types of coding and compression schemes that are workable within present day multimedia systems. Obviously, coding and compression will progress, and assuredly, some will give way to other methods and techniques in the years to come. MPEG and JPEG are the current standards for narrowband support of broadband requirements, while high-quality audio transmission requires 32 kHz bandwidth. The reader should be aware that not all coding and compression schemes preserve the full integrity of the signal(s) being conveyed. Methods involving some amount of signal degradation, referred to as loss techniques, may allow greater compression, but suffer from loss of signal definition.

References

(1)   McGrew, P.C., *On-Line Text Management: Hypertext and other Techniques*, Intertext Publications, McGraw-Hill, New York, 1989, pp. 40-50.

(2)   Goyal, S. K., and R. W. Worrest, "Expert System Applications to Network Management," in *Expert System Applications to Telecommunications*, J. Liebowitz (ed.), John Wiley, New York, 1988, p. 1.

(3)   Harris, C. J., and I. White (eds.), *Advances in Command, Control and Communication Systems*, Peter Peregrinus, Ltd., London, 1987.

(4)   Henning, W., "Bus Systems," *Sensors and Actuators A*, 25-27 (1991), pp. 109-113.

(5)   Hoffman, M., "Technology Profile Neural Networks," *Techmonitoring*, SRI International, July 1991.

(6)   Lemmon, A., *Marvel - A Knowledge-Based Planning System*, GTE Laboratories, Internal Report, 1986.

(7)   Martin, J., *Design and Strategy for Distributed Data Processing*, Prentice Hall, Englewood Cliffs, NJ, 1981.

(8)   Avedon, D., J. R. Levy, *Electronic Imaging Systems*, McGraw-Hill, New York, 1994.

(9)   Yager, Tom. *The Multimedia Production Handbook for the PC, Macintosh, and Amiga*, Harcourt Brace, New York, 1993.

Chapter

8

Applications Requirements for Multimedia Networks

Chapter Highlights:

Generically, multimedia networks deal with only a limited number of applications. Among these are networks that are used to allow individuals and groups of individuals to interact with one another between remote locations. This form of communication and interaction is currently referred to as video teleconferencing. If this application is solely between 2 individuals, the application is referred to as video telephone. In other applications, the actual working material must also be made available to distant partners simultaneously. This working material is typically found in some sort of database environment. Such requirements for information must allow access to databases, pictures, graphics, diagrams, text, and audio sources. Additionally, both or several parties must be allowed to view and discuss these source materials with the other without regard to where the source materials are located. This capability is referred to by several names, such as conferencing software. Third, individuals may sometimes want to access information available in several forms, such as through voice/sound, print, and video. This requirement can be provided through a series of storage media that can handle audio and images as well as textual data. This chapter will explore the hardware and software component requirements needed to support these applications.

8.1   General Requirements

The general applications requirements for multimedia platforms and networks
are straightforward. There are, however, a series of steps that have to be taken
in order to determine exactly what these requirements are [1:40-50].   For
instance, we have to define what the multimedia operation is, what its principal
features and functions are, and what range of capabilities is required in order to
finally determine the nature of the requirements.

The major effort and key ingredient that influences multimedia systems is
integration. Integration is the assembly and operation of the various elements
of a system that create the multimedia environment.   For instance, audio
collection and playback facilities that are combined with text or word
processing capabilities comprise one instance of a multimedia facility.

Within this facility operation, however, there are a set of features and
functions that dictate the response of that system to the requirements of the
multimedia system operation. The principal features and functions that make
the system work are as follows:

First, there are the delay functions between the parsing of the
various media segments and the reception and use of those media
segments at the receiving end. The media segment may be an image
frame, a speech clip, or a grouping of text characters.

Second, there is the quality of the media segment that is
transmitted over the system interface to the receiving end of that system.
The quality feature is the preservation of the information that was
initially transmitted by the originator to the receiver.

There are other features and functions that support a multimedia
network. Indicated below, however, is a top-level list of those capabilities that
are basic to all such systems and from which essential features and functions are
derived. They are summarized as follows.

(1)   The system(s) must be capable of utilizing a range of file formats in
order to provide ubiquitous usage and utility.

(2)   The hardware must be capable of providing input and output of
the principal media types, e.g., voice, print, and video.

(3)   The system and the network must be capable of storing a large
amount of data and media files of all types.

(4)   The system must be capable of accommodating many different applications programs commensurate with the hardware features available.

(5)   The system must be able to assess its own output quality and to provide that information to a network manager.

(6)   The system must be capable of guaranteeing a certain level of delay for the information delivery so as to assure that the integration of information at the receiving end will provide cohesive and uninterrupted reconstruction and presentation.

(7)   The system must have a network management architecture that links all elements of the operational system physically and functionally.

Each of these points will be discussed in more detail below. The range of applications programs available for multimedia may include not only the expected video and animation features thought to be part of the typical multimedia system, but also such additional capabilities as the integration of telephone, FAX, on-line data services, presentation software, etc. This range of intricate and complicated features makes the assessment of production quality even more important than would be the case with ordinary data networks. The network manager is unsuitable for such assessments because of the variability products generated, therefore, the applications programs themselves must be the agents for quality reporting.

8.2   Components of Multimedia Networks

As mentioned in the chapter highlights, there are about 3 major scenarios, or applications, that justify the use of multimedia systems, these being video teleconferencing, data, video and/or still image conferencing between parties such as a telephone conversation, and lastly, resource server utilization for informational, educational, or business applications similar in nature to a database access [3]. The identification of the situations in which these applications are to operate is important, because multimedia, by its very nature, is dependent upon its eventual use as a conversational aid. The addition of interactive capabilities has only increased the importance and utility of multimedia presentations. Currently, additional features are being developed, tested, and added to multimedia applications. Most of these new features are based upon interactive selections made by the user and presented by the application program. One such application will allow selective branching from one point in a presentation to another entry point based upon conditions expressed by the viewer, or based upon probability assignments, created by random number generators.

8.2.1   Multimedia Network Elements

Multimodal interaction between individuals or between individuals and information servers, acting as repositories for various types of information, is the major objective of multimedia.   The principal components of the multimedia network that supports these interactive scenarios are illustrated in Figure 8-1. The multimedia network consists of one or more data and/or media storage devices and the multimedia interface devices.

In Figure 8-1, the transmission system interfaces the storage servers, user interaction platforms, and the network manager for the system. The major applications software modules necessary for these operations are the database routines, the multimedia interface routines, and the network manager interface modules, all of which will be discussed below. One of the characteristics of multimedia interface routines is that they must be capable of handling many different protocols.

There are several problems associated with the implementation of multimedia network applications that should be discussed at this point.   In summary, these issues revolve around the large numbers of formats and protocols associated with different types of information files, and the variability in the number of equipment vendors. For instance, Figure 8-1 shows that each multimedia platform has its own set of applications and interface routines. And the applications modules associated with each platform may have their own special access and software updating features, depending upon the dissimilarities associated with the various protocols and vendors that provide the system connectivity. The capabilities of the different functions concerned, may require many different multifunctional accesses, interfaces, and manipulation routines to be incorporated into the system. This will ensure that the platform has ubiquitous utility and accessibility from most all other functions and vendor products with which the user software may wish to interact.

There are 3 principal components of the typical multimedia system. These include the server, the transmission system or network involved, and the appropriate workstation or platform. Some of the issues surrounding the server have been discussed above, while the platform and the transmission components have been introduced in previous chapters.   These principal components are outlined below.

Multimedia Server -   The multimedia server, or database, is the repository
                               for most of the multimodal pieces of information
                               needed for human-initiated interactions. These
                               database items include the applications programs,
                               such as teleconferencing software, animation, and

CD players, and the multimedia source files themselves. Since various media are stored on the same device or associated between devices while a multimedia presentation is being played back, each must have tags so as to associate different media files with each other.

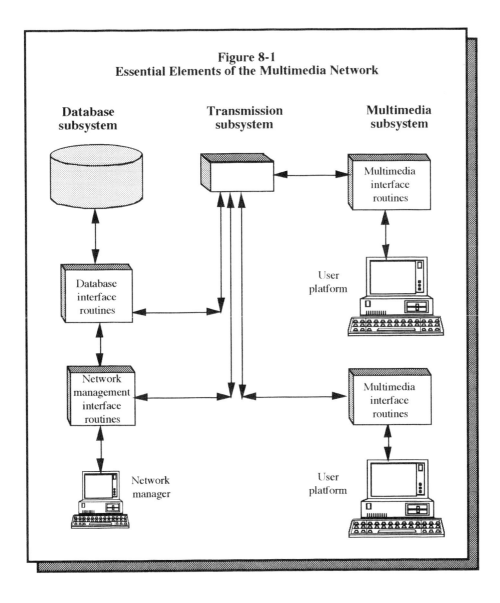

Figure 8-1
Essential Elements of the Multimedia Network

Database subsystem

Transmission subsystem

Multimedia subsystem

Transmission System - The transmission medium is the Achilles heel and principal bottleneck for multimedia, as with any data network. Bandwidth is usually scarce and at a premium for these types of applications. Copper is the basic fare for current communications and has severe bandwidth restrictions even without the loading coils commonly used in the telephone system to limit bandwidth. As cited in Chapter 2, many attempts have been made to extend the usefulness of copper in the wideband environments of today by developing compression techniques that can extend its bandwidth capability. Other transmission media, such as fiber optics, hold promise for future usage as certain drawbacks are overcome, but bandwidth usage will likely always require compression no matter what the bandwidth offered by this medium.

Multimedia Platform -   The media interface for the human communicator has to have input and output facilities for all of the 3 media types, as well as processing and interpretation capabilities for each. The electronic digital environment of today requires that a sophisticated computer system be available for each node to mediate the capabilities involved. The platform also has to provide all of the applications that can generate and playback the presentations created.

8.2.2   Multimedia Network Protocols

A multimedia network may consist of several platforms with different access requirements. An example of a multiplatform access requirement would be a multimedia animation player that has to access information files from other platforms in order to conduct its animation scenario.   In such a case, a dissimilarity of the application software, its communications interfaces, between the player software and the database software would require protocol or access conversion routines to be made available at both ends of the network in order to assure or facilitate communications and the access of necessary applications files.

On the surface, the problem of finding the correct set of data files on a network, as described in the previous paragraph, is not much different from any other data communications interface issue. However, the variety of conversion requirements and media access standards necessitate many more access possibilities than would be the case with an ordinary data communications network. As an example, just the number of image formatting standards alone adds up to well over 30 at present. Thus, the transmission of an image between 2 multimedia platforms may necessitate access to at least 2 different storage formats out of many dozens of possibilities, in order to be assured that the proper decoding and encoding of transferred images is available to the multimedia network.

Finally, traditional database storage protocols have the problem of requiring a considerable amount of overhead in the form of directory and accounting information for too little data per storage location to be suitable for pictures, video, or audio clips. Such formats as GIF, PICT, and TIFF were developed to solve just these objections, since they actually compress the data. Some of these algorithms do not incur information losses, evidenced as noise or other distortions in the image space, while other algorithms actually destroy some of the information content, such that the decompressed images show increased noise, smearing, streaks, or other image corruption.

Thus, an appropriate interface in the form of directory and access support must also be developed to support such features as well. Consider Figure 8-2. Any multimedia or image file has its own overhead requirements based upon the file headers and data segmentation format of the file standard. In addition, any storage device, such as a magnetic disk, also has its overhead, or segmentation requirements and directory reserve area on the storage disk needed to identify files on the disk and their location. If these overhead requirements could be streamlined, more information could be stored per storage unit with more efficient location and retrieval mechanisms available.

Figure 8-2 illustrates differing formats using special icons sometimes accompanied by descriptive abbreviations. Internally, these icons are associated with the particular image or video file via links to the file directory information. These linkages are difficult to compress further, thus, the storage advantage must come in the compression process itself.

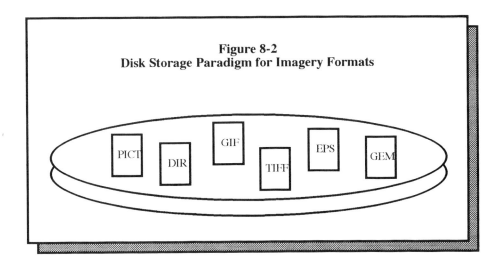

Figure 8-2
Disk Storage Paradigm for Imagery Formats

Each of the components outlined above, especially the user platform and the network manager platform, has its own unique characteristics, so a unifying software capability has to be provided to format all outgoing requests and/or media files, and to interpret all incoming messages and media files. The media platform software must be able to compress or decompress incoming or outgoing data sets as well as sample the data streams for messages, destination addresses, and the like.

8.2.3   Applications Packages

The applications associated with multimedia are more variable than the typical word processing program in that multimedia packages provide a range of sensory presentations. Additionally, each applications package has its own set of nominal operating characteristics. That is, each has its own operational range within which it best performs or can perform. If the data handling requirements, such as sampling, coding, transformation, presentation, manipulation, etc., alter the application program's ability to service the current jobs at the transmission system's required service rate(s), then data problems will begin to occur. The inputs, with their required rates, coupled with the limitations of the transmission subsystem, dictate that the processing subsystem alter its processing to accommodate these 2 constraints. This potential processing problem creates a possible overload situation.

In the digital processing environment, video distortions appear in the visual presentation as ghosting, smearing, jerkiness, etc., while video processing overloads materialize as frame dropouts, noisy images, image frame overlaps, and the like. For example, the North American TV signal standard, known as NTSC, calls for data frame transmission at 30 frames per second. Each frame duration equates to $1/30$, or about 33 milliseconds per frame. If any or a combination of any of the above-mentioned video processing tasks requires more than 33 $ms/frame$, including system overhead, degradation of the processing capability becomes evident, unless the source data is first stored and later processed for transmission. The European standard, CCIR, has a 25-$frame/s$ specification, which translates to an upper limit on real-time processing of 40 $ms/frame$ including system overhead. Thus, each frame exercised on the CCIR standard has a dwell time of 40 ms on the display screen before the next frame is displayed. On the other hand, video presentation distortions are more a result of bandwidth, interference, and other corrupting influences.

Video processing problems result from insufficient time to retrieve, decompress, decode, or otherwise reformat the image frames for display within the allocated times. With regard to the problem of compression and storage, many different methods have been devised to reduce the bandwidth necessary to transmit video frames. The International Standards Organization (ISO) video compression algorithms known as the MPEG series are part of this array of image processing algorithms. The data processing requirements, at least as far as the MPEG standard is concerned, go far beyond the real-time processing time constraints of either a 33- or 40-ms time limit, and therefore the data must be stored, processed, and transmitted at some point later, i.e., in non-real time. The MPEG processing, as an example, may require up to 10 times the 33-ms constraint per frame. Thus, processing of a feature-length movie may require over 20 hours for a 2-hour movie. This clearly is an unacceptable constraint for narrowband image transmission. Current tests on MPEG have confirmed this lengthy processing time frame.

The processing of video frames may also require several steps in order for the complete compression and/or enhancement process to be realized. Currently, most applications routines for video proceed through a set protocol of algorithms. These applications routines do not consider the time constraints of the total processing activity, the alteration of steps that may be required if the processing takes more time than planned, nor the reporting of status to a network monitoring function. With regard to the first issue, the frame rate may have to be adjusted to keep pace with the processing requirements. Second, the dynamic activity of the processing steps may have to be adjusted in order to fit within time constraints; and third, each alternation of the processing should be reported to a monitoring system.

In the NTSC system, for each frame that is dropped, these reduced frames will increase the overall dwell time for each frame and therefore the processing time available for each cycle. For each processing cycle, the algorithm may have to be adjustable in order to fit within the display cycle required. The objective of efficient multimedia network management is not totally accomplished without the network manager having oversight over the quality of the transmitted multimedia information [2]. In order for effective network control to be realized, comparison criteria must be established first.

8.2.4   Network Management Implications

The monitoring system addressed above is one part of the overall network management system. Probably the most important ingredient of the network management system in the multimedia environment is its ability to pass status-of-quality information associated with the applications being exercised. The applications program should not only be capable of assessing and passing status results about the information being generated, but also should have an ability to detect and report on its inability to keep up with data processing needs. This set of requirements is depicted in Figure 8-3.

Figure 8-3 shows how all of the issues discussed so far are integrated onto a multimedia applications platform. The multimedia system is depicted by the computer icon, which in turn has a set of applications programs as part of its repertoire of capabilities. These programs may include animation, teleconferencing, interactive CD features, and other interactive information sharing routines. These applications features are routed to either external communications interfaces or back to internal features, such as database storage.

The overall process of managing a multimedia network is complicated even more by the fact that there are different modalities, namely audio, print, as well as video. Video is elaborated upon here because of its considerable bandwidth needs as opposed to audio or text.

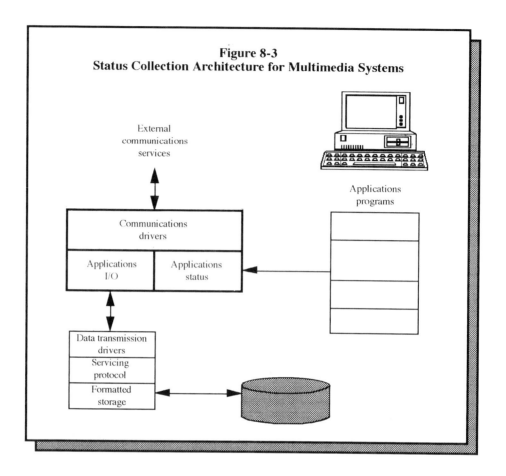

Figure 8-3
Status Collection Architecture for Multimedia Systems

As indicated above, criteria must be established in order for the comparison of normal quality standards with altered ones to be made and evaluated for acceptability. If one of the criteria is noise analysis, then the following comparison might be performed.

First, consider a "normal" or otherwise acceptable image stream that contains an acceptable or known noise content. This noise level could be tantamount to the image illustrated in Figure 8-4A. This image has some noise content to be sure, but the image is clear enough to identify the objects in the scene with considerable clarity.

Figure 8-4B is a histogram of the pixel values within this image. Notice the periodic spikes of pixel values throughout the range of the histogram. These spikes are indicative of a relatively high contrast, or sharp difference, between adjacent pixel locations and neighboring ones. These sharp spikes are not expected to be arranged in any particular distribution, but rather portray a picture of pixel value ranges that accompany an image of high contrast and, hopefully, clarity.

Figure 8-4A
Video Frame

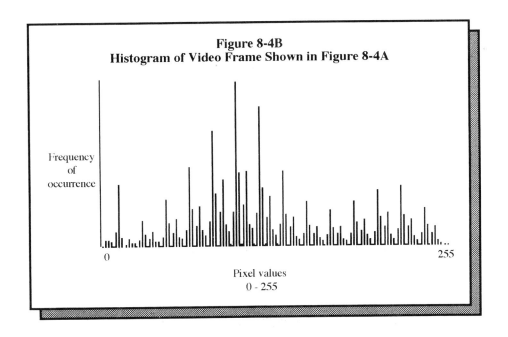

Figure 8-4B
Histogram of Video Frame Shown in Figure 8-4A

Frequency
of
occurrence

0   255

Pixel values
0 - 255

Figure 8-4C
Noisy Video Frame

Figure 8-4D
Histogram of Video Frame Shown in Figure 8-4C

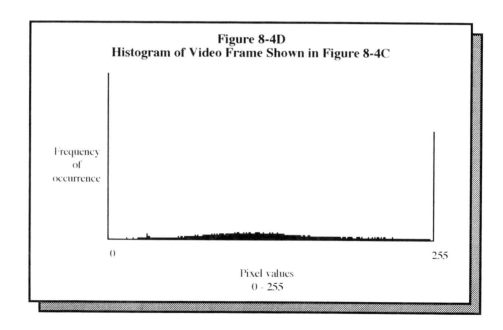

Frequency
of
occurrence

0   255

Pixel values
0 - 255

Figure 8-4E
Abstract Objects

Figure 8-4F
Abstract Objects Histogram

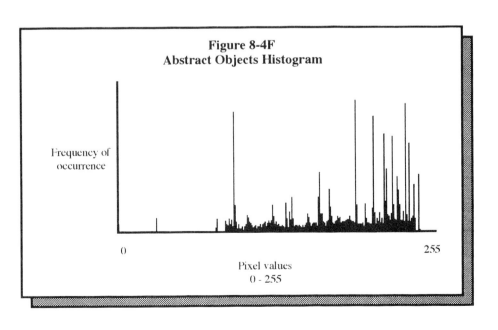

Frequency of
occurrence

0   255

Pixel values
0 - 255

Figure 8-4G
Corrupted Abstract Objects

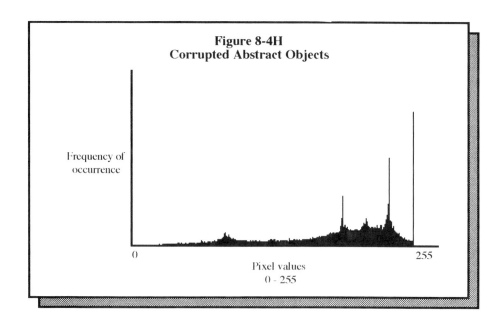

Figure 8-4H
Corrupted Abstract Objects

As seen on the histogram in Figure 8-4B, the pixel values are displayed on the horizontal or X axis, while the frequencies for each pixel value are displayed on the vertical or Y axis. The pixel values range between 0 and 255 as the image has a gray-scale depth of 8 bits. The figure shows that there is a very definite distribution of pixel values within that particular image. A noisy image would have had a more even distribution of pixel values across the acceptable range of values.

Figure 8-4C shows the same video frame as depicted in Figure 8-4A, but corrupted with a considerable amount of noise. It is obvious that this frame contains noise because of its grainy appearance and fuzzy display of certain objects within the frame. Figure 8-4D is a histogram of the image shown in Figure 8-4C where the distribution of pixel values within Figure 8-4C are portrayed graphically. Notice that this histogram portrays the pixel values as being spread over a wide range of values, but no set of pixel values has a particular frequency range associated with it.

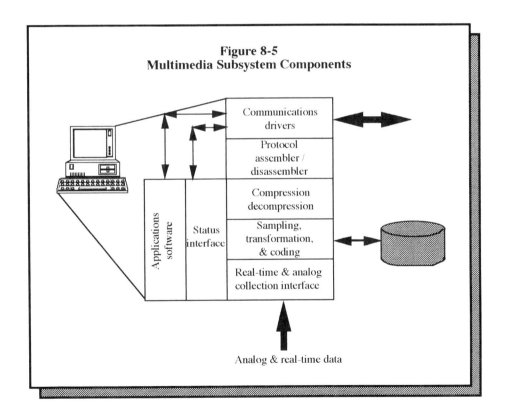

Figure 8-5
Multimedia Subsystem Components

Thus, the histogram for the noisy image has fairly constant frequencies associated with each pixel value, much the same as would be seen in a white noise display. Notice that there is a wide range of pixel values in Figure 8-4D across the 8-bit range from 0 to 255. Theoretically, true white, or broadband, noise would be shown as an even distribution of pixel values across the entire range of possible values. This theoretical distribution is close to realization in the actual portrayal shown in the histogram figure.

Now, we will take a look at the opposite process. Figure 8-4E shows a computer-generated group of objects uncorrupted with noise. Notice that Figure 8-4F shows the histogram of the pixel values of this image. This histogram is similar to that representing the image in Figure 8-4A. Figures 8-4G and 8-4H show the same abstract objects after being corrupted by noise as well as the accompanying histogram, respectively. Notice also that the histogram in Figure 8-4H is similar in the spread of pixel values to that represented in Figure 8-4D.

The total multimedia system components are depicted in Figure 8-5. The set of components in the multimedia system can be summarized as being a complement of communications servers, protocol converters, compression capabilities, and analog-to-digital conversion software. Each such platform should have all the facilities indicated in this figure in order to be complete and sufficient for multimedia work.

8.2.5   Multimedia Information Loading

The imagery associated with multimedia has a wide range of storage needs. For example, the typical uncompressed FAX generates approximately 1730 picture elements (pixels) per line and there are about 100 scan lines per inch. A typical sheet of paper requires 850 scan lines times 1730 pixels per line, or about 1.4 MB of data. The equivalent number of bytes would be approximately 184 kB per page. Equivalent imagery would consume the following volumes in the formats designated:

Format	Storage Volume (Bytes)
GIF	69k
PICT	120k
TIFF	123k
FAX	184k

FAX can be highly compressed, and with compression ratios of 8:1 or better, its storage requirements are about 23 kB as compared to the image formatting schemes that are between 3 and 6 times this value. Text information has the least volume requirement of any of the 3 major media modalities. A full sheet of text requires about 500 characters. At 8 bits per character, the resulting storage requirement typically is the 500 bytes of information plus an overhead of special characters, to include formatting, that increases the total byte storage to 3.5 times the information content, or about 1800 bytes.

Audio information, on the other hand, can be sampled and converted to digital format at differing sampling rates. The minimum acceptable frequency for human voice is about 3 kHz which means that the sampling rate has to be at least twice that, or about 6 kHz, using the Nyquist criterion of 2.2 times the analog frequency to derive the digital sampling frequency. Other criteria for audio sampling are 32 kHz for a high-fidelity analog signal of 16 kHz, and a maximum sampling rate of 44 kHz for an analog signal of 22 kHz. Compression algorithms would reduce the storage requirements of any sample accordingly. The following table shows the uncompressed storage requirements for several sampling scenarios:

Sampling Rate (bytes/s)	Storage Volume (bytes)
6k	30k
32k	160k
44k	220k

8.3   Multimedia Data

The requirements for multimedia applications are defined and constrained by the data involved. These issues will be discussed in this section. The major items of interest in this area are the categorization and organization of multimedia data sets.

Multimedia data comes in 3 major varieties, text, audio, and imagery. Imagery includes video frames, still images, scanned pictures, drawings such as art work, and graphic designs. Each category will now be reviewed briefly. FAX can be considered a hybrid between imagery and text. Originally, FAX was a medium for the conveyance of text. Technically, however, FAX is an image transmission medium. More recently, software has been developed to resolve FAX images into the text material and other images that are contained within the FAX image space.

8.3.1   Imagery

Some of the early pioneers in the arena of image formatting and storage included Compuserve and Apple. Even during the early days of on-line information transfer, at a point when text information sharing was at the forefront of information technology, Compuserve recognized the need to establish a standard for the formatting and storage of still imagery. That company developed Graphics Interchange Format (GIF) to address this problem. This format is not quite as efficient as others that have been created since then, but it has worked well for the environment in which it has operated.

Later, Apple developed a file format named Macpaint that was its initial entry into the file format environment. This format produced small files using compression techniques, and was suited for black-and-white imagery and for images that were to be printed on laser printers of 300 dots per inch (DPI) densities. Later Apple developed other formats, such as Picture Format (PICT), that further improved the compaction of the image data beyond Macpaint and other formats that evolved later. Other formats developed and frequently used today include Tagged Image File Format (TIFF), and Encapsulated Post-Script (EPS).

Graphical/pictorial data storage systems, on the other hand, are relatively new technologies. They begin with the storage formats used for pictorial data. Some of these include PICT, TIFF, and others developed mainly by Apple or third-party providers who originally catered to Apple applications. These image storage capabilities are formats that were created to allow images to be recognized by different software application packages.

The nature of multimedia is that there are many different storage formats for each type of media involved. As mentioned previously, there are many formats for the various text applications, several types for the audio clips that may be required, and many varieties of formats for imagery and video. There are even formats for animation. Given this, the storage of multimedia programming will require different formats, and therefore different files, in order to deliver a media presentation.

8.3.2   Text

Text processing began with the advent of the Morse code at a time when words or symbols were represented by groupings of alternating transitions in the electrical signaling, sometimes referred to as dots and dashes. This system survives today, as it is used frequently in marine and sometimes in air-to-ground communications. These coded representations are of variable length, and while efficient for the intended use, they are not appropriate for current-day usage because of the limitations of characters and symbols. The next innovation was a system known as binary coded decimal (BCD), and later, IBM developed a coding system known as extended binary coded decimal (EBCDIC) code for use in digital computers.

Data storage systems for textual information have received much more development attention than those for graphics, still frames, or video. These data storage systems are almost too numerous to review, but the best-known ones include:

(1)   **IMS** -

This is a hierarchical database retrieval software package that was used on large computer main frames. Hierarchical database management systems are relatively fast for tree search types of applications, but slow for category types of searches. Images are stored in compressed or uncompressed files, and are inherently tree structured in nature. However, header information in each image file may contain reference information that can be used to perform category searches. Such information might include time and date stamps, and key word inclusions provided by the author.

(2) **Oracle -**

This database capability is a relational database management system originally developed for large computer systems, and later adapted for networked workstations and eventually PC networks. Others involved in relational database systems also include Ingres.

(3) **SQL/DS -**

SQL/DS is yet another relational database management system that has become a standard against which other data query systems have been measured. SQL, or something like it in functionality, can be valuable in both hierarchical and relational image searches.

8.3.3   Audio

Sound is treated in much the same way as text files or imagery. First, sound documents can be created using sound import capabilities or by making use of preexisting sound documents from which sound clips may be taken and integrated into other sound document creation file capabilities. Second, sound resources, as mentioned previously, can be imported or supplied as existing sound source files. Third, these files are formatted using their own format algorithms, one of the most popular being the Audio Interchange File Format (AIFF). These 3 elements of the usage of sound define the total of this important medium's utilization within the realm of multimedia. The sound clip, or clips, used in some particular application are usually stored within the application within which it is to be used. An example would be the storage and recall of a sound clip within a Microsoft Word file. Sound clips can be stored in their own folders for later usage in whatever applications are deemed appropriate. In both instances, the AIFF format, or some other, can be used to format the audio for playback as needed.

8.4   Issues Related to Multimedia Applications

8.4.1   Multimedia Data Transmission

Multimedia applications consume considerably more bandwidth than is required by data traffic. The typical video picture frame requires, at a minimum, 8 bits per pixel, times 460 by 320 pixels, or 147,200 bytes per frame. A typical page of text requires about 2400 bytes. The bandwidth expansion requirements are over 60:1 for video over text. While a frame of video is being transmitted over a wideband link, other information requests may also need to be honored. If these requests cannot be so honored, a continuous stream of video and audio may not be possible. As a result, video is transmitted in packets rather than in frame-sized chunks.

These packets are variously constituted depending upon the protocol used. For example, the ATM protocol uses 48 bytes of information preceded by a 5-byte header. This means that a video frame carried over an ATM network requires 3067 packets to transmit that one picture frame. In order to maintain a 30-frame/s rate for a typical video presentation, the transmission rate has to be at least 39.4 Mb/s, uncompressed. Additionally, each packet has to be delivered to the transmission interface at a fairly constant rate in order for the aggregate number of packets to be reconstituted and displayed at the receiving end. This strobe rate has to be about one packet every 10 μs.

8.4.2   Multimedia Data Reception

The next major issue has to do with the variability of the arrival rate of packets at the receiving end. At the receiving end, the collection and reassembly of the transmitted packets, whether they be packet audio or packet video, have to occur in a fairly regular manner in order to maintain a relatively even flow of data and subsequent usage of that data. There is some but not a 100 percent correlation between the rate and variability of transmission and the rate and variability of reception of that data. As a practical matter, the reconstruction of each frame may have to be delayed for some constant time, such as having a forced delay of one full frame or 33 ms, or $1/30$th of a second. If the standard deviation for each received frame packet set is, say, 10 ms, then in order to assure a full set of pixel values with a 5-standard-deviation coverage, any frame could require 33 ms plus an extra 10 ms in order to transfer its full complement of values or 83 ms. This means that buffer space for at least 3 frames must be made available in order to assure continuous reception of frame pixels for any particular frame. Technical Note 8-1 illustrates the variability problem for the frame collection problem.

Another set of problems comes to the surface when the issue of delays is discussed. These problems revolve around the subject of identifying which frame any received packet containing pixel information pertains to. This identification problem must be solved within the image transfer software because the ATM platform is oriented to the routing of information, not its proper identification within the data stream. Thus, the number of packets will have to be increased in order to accommodate such pixel-frame tracking overhead. The second problem is how the packets are being routed through the system. If they are forwarded directly from sender to receiver, there will probably be no variability, but if there are 2 or more routes that may be taken by the packet routing subsystem, then the complications and variability of reception are increased. Technical Note 8-2 illustrates this paradigm. Pathway d1 is direct, and if each packet is strobed out to the communications interface without internal delays, then each packet is received in order at the receiving end and can be unpacked, decoded, and stored in the appropriate display buffer. Pathway d2 is indirect and is subject to delays on both links plus additional delays associated with receipt and retransmission at the intermediate point. Thus, synchronization for pathway d2 is very difficult.

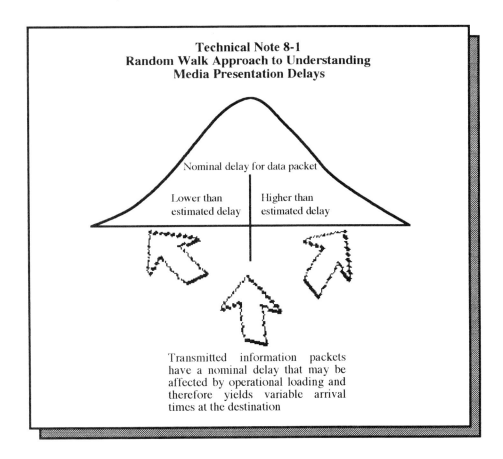

Technical Note 8-1
Random Walk Approach to Understanding
Media Presentation Delays

Nominal delay for data packet

Lower than estimated delay

Higher than estimated delay

Transmitted information packets have a nominal delay that may be affected by operational loading and therefore yields variable arrival times at the destination

The acceptable solution to the probability that a transmitted picture frame will not be perfectly synchronized is to accept some range of delays such that 2 or more buffers will be used to store specific arriving frame pixels for display at the proper times.

8.4.3   Quality Determination

Quality determination has been discussed in previous chapters, but is reintroduced here for completeness. The take-home message here is that multimedia information transfer is not only a matter of accurate data interchange, but also an issue of the quality of that information transfer. Again, the difference between a clear picture and the same picture corrupted by transmission errors or noise may make a difference as to how that picture is received psychophysically by the human receiver. Issues of credibility and accuracy may slant the viewer's interpretation of the visual or auditory results.

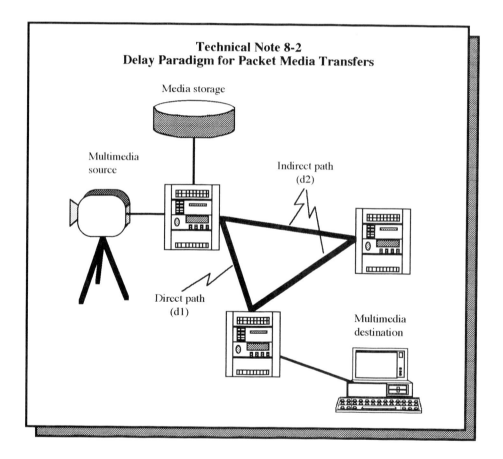

Technical Note 8-2
Delay Paradigm for Packet Media Transfers

The quality measurements made by the sending system can be passed to the receiver and/or the network management system so that trends in quality can be tracked and acted upon as required to assure the best possible conveyance of multimedia information from one point to another. Figure 8-6 shows the points at which quality information is created and can be used for evaluation. If compression algorithms are lossless, their decompression will not affect the quality measures, but if the opposite is true, then the quality measures may need to include such indications so that the viewer can be warned about such possible corruption.

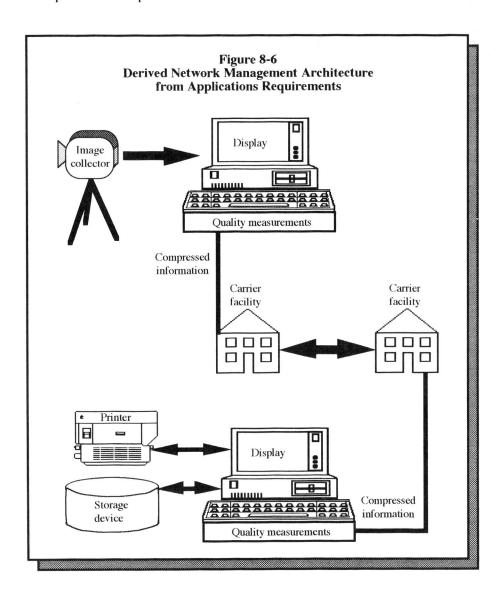

Figure 8-6
Derived Network Management Architecture
from Applications Requirements

Summary

There is a tight coupling between the multimedia network management system
or systems and the applications packages that operate on these networks. The
generalized requirements that were outlined in the first section rely, mainly,
upon specialized and well-packaged hardware features that allow the user to
collect and then to output media information over the supporting network to
the intended users.   The rest of the chapter explores the categories and
organization of various categories of multimedia data. The principal network
management feature discussed is the information assessment tool that relies
upon the continuous comparison of information as the platform outputs
multimedia information over the network. As an application requirement, this
measurement tool is key to an effective network management capability for any
multimedia network.

References

(1)   McGrew, P. C., *On-Line Text Management: Hypertext and Other Techniques*, Intertext Publications, McGraw-Hill, New York, 1989, pp. 40-50.

(2)   Goyal, S. K., and R. W. Worrest, "Expert System Applications to Network Management," in *Expert System Applications to Telecommunications*, J. Liebowitz (ed.), John Wiley, New York, 1988, p. 1.

(3)   Harris, C. J., and I. White (eds.), *Advances in Command, Control and Communication Systems*, Peter Peregrinus, London, 1987.

Chapter

9

Multimedia Network Control Techniques

Chapter Highlights:

This chapter proposes an architecture for the assembly and delivery of multimedia information that will support the network management function. This process is referred to here as a quality tag, which is generated periodically by the collection platform, reassessed at the receiving platform, and compared by one of the subsystems of the overall network management system, known as the multimedia node of the network management system. This quality tagging and evaluation process is developed stage by stage in this chapter with successively more detailed diagrams so that the reader can assess the usage of this technique at each stage of the operational activities within the multimedia environment. The development of a quality tagging system is, on the one hand, difficult because of the various parameters involved. On the other hand, it is important because of the ramifications associated with the delivery of various media for a focused and concerted purpose.

9.1   Network Control Criteria

There are several criteria that must be applied to the control and eventual management of multimedia network environments. As described in previous chapters, voice and data network management has heretofore consisted of the monitoring and control of those network elements up to and including levels 4 and 5 of the Open Systems Interconnect (OSI) model. Levels 6 and 7 have been assigned to the operating systems and the applications programs of the software involved. The reason for the limited attention in data networks only to level 5 was either that there were no presentation nor applications standards for network management usage, or that such standards were not useful for network management. Another major reason was that the level 6 and 7 infrastructure upon which these networks operated either did not exist, as in the case of purely analog voice, or the efficacy of the applications depended upon the physical or functional integrity of the network elements at the lower levels of the model.

Network monitoring parameters, such as packet error rates, appear to afford little multimedia management value, because these parameters are directly dependent upon the network elements rather than the applications themselves. In other words, network management in the past has focused upon the integrity and efficacy of the transmission system. The management of multimedia networks is concerned with the presence and quality of the applications involved as opposed to their physical capabilities.

In addition, the principal objective of data network management is to isolate problems once they have been identified, or to predict such potential problems to the extent possible. As such, error rate information is only one of many sources of monitoring and says little about the possible root of the situation. The principal objective of multimedia networks is to predict, identify, and eliminate sources of problems in the applications of the media involved. Thus, measures of quality in the perception of the information are much more important than in data network management, where error rates are considered quality measures.

Furthermore, network management capabilities in data networks rely principally upon the hardware comprising the network. The functional features of these hardware devices consist of the packet and switching logic necessary for these devices to have utility. These features include, packet assemblers and dissemblers (PADs), smart T-1 or T-3 multiplexers (muxs), packet and circuit switching equipment, and the like. Each of these has its own monitoring features that can be accessed remotely from monitoring and control centers. A high packet error rate alerts the controller to a potential problem that must be further investigated. The idea behind data transmission systems is that they are essentially bimodal, i.e., the information within these packets is either correct or incorrect. There are no shades of gray, nor partially correct answers.

The software within these network elements acts only to isolate correct versus incorrect situations within the confines of each network element and not across the network itself. Additional control requirements are imposed by the software functionalities inherent in such network features as packet protocols of various types are monitored. In multimedia systems there is, obviously, heavy reliance upon hardware, but the contents of the data packets that carry the information, rather than the wrappers, is the major concern for this information.

Therefore, the quality of multimedia system operations is affected not only by the degree of network management control of the physical and functional transmission elements, but also by the quality of the actual communications. In terms of quality, these actual communications are affected by both the other modalities involved as well as the attributes of the applications generating their transmissions [3]. In other words, multimedia networks have complicated and diverse network elements that support the 3 principal modalities involved in the conversational interactions. As such, the conveyance of sensory information is not a trivial activity because of the complexity of the technologies involved. In addition, the management of multimedia networks that support these applications must also monitor, evaluate, and control the quality of the information.

9.1.1   Quality Tagging

Each picture image, video clip, text passage, or graphic display must give an indication of its quality in order for the multimedia node of the network management system to assess its quality at the receiving end of the transmission path [7]. The network management system will have several submanagers, possibly one for each of the main subsystems, such as the T-1/T-3 backbones, the packet assemblers and dissemblers (PADs), etc. This concept is illustrated in Figure 9-1. Subsequent sections will address the breakdown of this concept as depicted in this figure.

Figure 9-1 shows the three main components of multimedia, voice, print, and video, being generated by their own unique hardware and software functional elements feeding into an integration controller node. at this point each mode is examined and a quality tag is assigned to that segment of the input. A segment may be defined as a period of time or a functional breakpoint. A functional breakpoint may be the duration of a text message, a data file, a still image frame, a video clip, a voice utterance from one person involved in a 2-way conversation, etc. [7]. A set period of time may be defined arbitrarily as 1 s, 100 ms, etc.

The tagged information is then sent on its way to the receiving multimedia platform for conversion to display form and quality analysis at that end. The received quality checks are based upon the same parameters as those generated at the transmission end, but additional features may need to be examined, such as incoming information rates that may necessitate buffering, and display control so that information can be strobed to the output hardware at a rate and in a manner consistent with the requirements of those media elements.

Implicit within this conceptual diagram is the need for some point at which the tags from both the transmission and receiving ends are compared for inconsistent quality results. If, for example, the transmission end generates a tag that negates any constraints upon the media collection equipment, but the receiving end generates a tag that identifies one or more quality issues associated with that information set, then the media submanager should be capable of identifying to the overall network manager that a problem exists and the conditions that have created that problem. This set of issues will be addressed in the next sections.

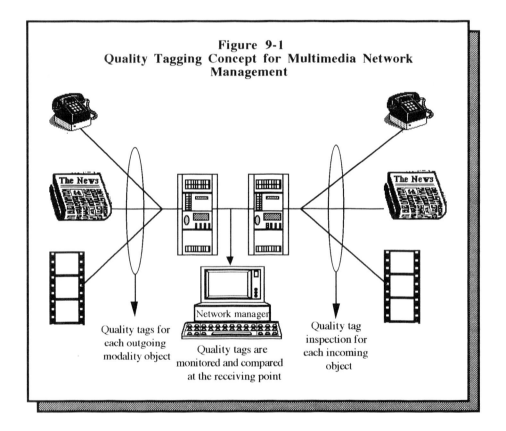

Figure 9-1
Quality Tagging Concept for Multimedia Network Management

9.1.2   Quality Requirements

Each modality has its own particular quality requirements, and each modality will be listed and discussed below.

(1)   **Video -**

Video is a flexible medium in that its quality is measured and
evaluated according to many different factors, some of which are
qualitative and some quantitative. The overall quality of a video
clip is a somewhat subjective judgment, and therefore the various
factors that are a part of this quality decision should be understood
in their quantitative respects. The received results can be judged to
be the result of a network management problem if such measures
are not available for comparison at both the sending and receiving
points, especially the receiving points. These factors include:

(a)   Image space size, i.e., pixels per line and lines per image area

(b)   Compression algorithms, if any, used in getting the image
        ready for transmission

(c)   The number of frames per second used in collecting or
        transmitting the imagery

(d)   Noise suppression or image enhancement algorithms, if any,
        used in cleaning up the images prior to transmission

The transmission of a set of frames composing a video clip can
many times also appear to have been improved by the transmission
process, due, in part, to the modification of that clip by altered
transmission requirements. For example, a noisy video clip can
appear to be vastly improved by the transmission of that image in
an altered format, such as fewer numbers of pixels in its transmitted
form.

(2)   **Audio -**

Audio has many characteristics that can be parameterized and
used for the quantitative evaluations of the quality and therefore
the network management value of these parameters. New
approaches to the packaging of these parameters, such as MIDI,
which will be discussed later, provide a means of consolidating all
of these criteria into a single choice or range of choices. The
parameters of interest to audio capture quality are outlined below:

(a)   The sampling rate of the audio signal

(b)   The depth of the sampling conversion, such as 8 bits per sample, 16 bits per sample, etc.

(c)   Multiple channels required for each sound

(d)   The analysis of each sound track, such as frequency analysis

(3)   Text -

Text is the oldest of the electronically transmitted data types, and is still the most common medium. It has been transmitted in various venues, such as one letter at a time, one word at a time, one message at a time, and one file at a time [7]. Its characteristics are such that it requires few quality checks compared to audio and video information, because it is either right or wrong without regard to subjective qualifications that one usually associates with images and audio information. The quality parameters associated with text material are [7]:

(a) Legal characters

(b) Error checking

(c) Information dropouts

At a slightly more detailed level of analysis, Figure 9-2 shows the linkages between the principal components of the multimedia network management architecture discussed above. The session interactions between the communicating nodes are processed and exchanged as dictated by operational protocols. Each burst of voice, print, or video information generates a quality tag that is routed to the appropriate network management node for analysis with the tag(s) received from the termination point(s). In cases where there are multiple reception points for multimedia bursts, the multimedia node may have to receive quality tagging reports from several different intermediate management nodes.

9.2   Error Reporting and Evaluation

This section will discuss, in overview form, some of the possible protocols, formats, and parameters associated with the quality measures discussed previously. This can be best organized by analyzing each of the parameters discussed above. Each of the parameters associated with the 3 principal multimedia modalities is elaborated upon below.

The whole objective behind the network management of any data or information system is error-free operation. Data networks operate asynchronously in that they are sometimes sending or receiving data and sometimes they are idle. Thus, any outages experienced by these systems are many times obscured by the periodicity of a system's activity. For example, if the system is not responding properly, this revelation may go unnoticed for some time and obstruct or hinder accurate data traffic, until the network manager becomes aware of the problem.

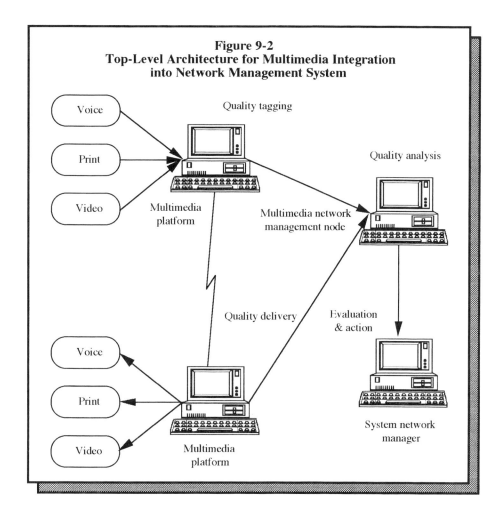

Figure 9-2
Top-Level Architecture for Multimedia Integration
into Network Management System

On the other hand, a problem with a multimedia interactive conversation will likely become very noticeable at the moment the problem occurs. An anomaly in an ongoing conversation involving voice, print, or video is easily determined by the human organism. Such errors in reception become evident through such measures as noise, distortion, fading, and the like. The nature of the human sensory systems is such that they are very sensitive to anomalies in their media.

The interpretation of the quality tagging concept is illustrated in Figure 9-3 using the scenario of an image scanning system. At the origination side, the system measures and assigns a set of quality values to the collected information. This, or these, value(s) are sent to a storage repository for later comparison at the received side of the transmission. At the receiving side, the transmitted image, text, and/or audio information may have been corrupted to some extent by the transformations and transmission relay handling. The received quality measures are again calculated and compared to the transmitted values. Such comparisons are used to assess whether the physical and/or processing elements should be modified.

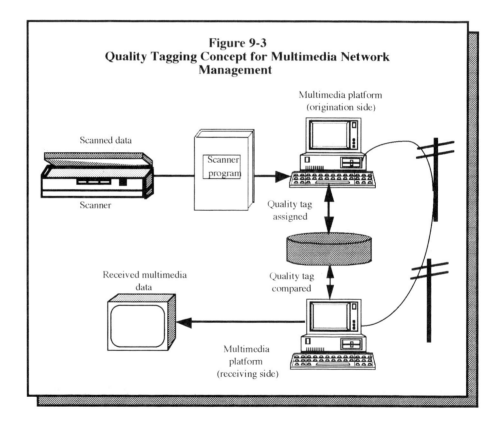

Figure 9-3
Quality Tagging Concept for Multimedia Network Management

9.2.1   Video

Video possesses several attributes that can be used to provide a measure of its displayed quality. These attributes represent the video's ability to faithfully convey the scene in as natural a presentation as possible. These attributes have been arranged below so as to maximize their multiplied results with maximal presentation quality. The ratios defined below are also arranged so that the higher the value for each ratio, the higher the quality of the output imagery.

Each ratio can be defined as follows:

(1) $\left(\dfrac{\text{XPD}}{\text{CPD}}\right)^2$ -

This is the ratio of the transmitted frame size divided by the collected frame size. The quality of the transmitted frame is preserved if it represents all of the collected pixels at the source. As the frame is compressed in size, so is its quality compromised. Hence, the ratio is squared to reflect the deterioration of the frame quality as the frame size is diminished.

(2) $\left(\dfrac{\text{Frame depth}}{16}\right)$ -

This is a value that reflects the depth of the gray-scale representation of the image or its color equivalent. Some images contain only 4 or 8 colors, or 2 or 3 bits of depth. The degree of color or gray scale gradation provides increased utility to the user, especially where graphics and imagery are concerned. Thus, the quality measure increases as the range of scaling increases.

(3) $\left(\dfrac{1}{\text{Compression ratio}}\right)$ -

The compression ratio is the measure of the degree to which the image has been compressed for transmission by a fewer number of bits than originally called for in the standard transmission format. Some compression algorithms can reproduce the original image without loss, while others will corrupt the image permanently. As a rule, some image degradation can be expected when compression is invoked. Thus, the factor illustrated above will reduce the image quality, and therefore quality measure, as the compression ratio increases.

(4) $\left(\dfrac{\text{Number of noisy pixels}}{\text{Total pixels}}\right)$ -

Another measure of the quality of an image is its original condition. This can be expressed as a ratio of the number of noisy pixels in the image space to the total number of pixels represented in that image prior to transformation and transmission. The greater the number of noisy pixels, the lower the quality to be perceived.

The quality measure that could be provided with the video burst is outlined as follows:

Video quality =

$$\left(\frac{\text{XPD}}{\text{CPD}}\right)^2 \left(\frac{\text{Frame depth}}{16}\right) \left(\frac{1}{\text{Compression ratio}}\right) \left(\frac{\text{Number of noisy pixels}}{\text{Total pixels}}\right)$$

where
XPD	= transmitted pixel density
CPD	= collected pixel density
Frame depth	= color or gray-level depth of image, e.g., 8 or 16 bits
Compression ratio	= compression ratio achieved with the compression algorithm utilized
Number of noisy pixels	= number of pixels having defined noise features
Total pixels	= total number of pixels shown in the displayable frame

9.2.1.1   Frame Size

Frame size is a quantitative feature of the displayable video motion. Typically, video frame sizes are displayed using an aspect ratio of 4:3, or 4 units across the screen horizontally for every 3 units of depth down the screen vertically. The larger the number of pixels in both directions, the bigger the total number of pixels that must be transmitted, or at least submitted to compression, and therefore, the wider the bandwidth required to support such an increased pixel space. The size of the frame determines a couple of parameters, one being the distance at which the viewer must see the image for best results of resolution. The other parameter relating to size is the clarity of the image presented.

9.2.1.2   Frame Depth

Frame depth is an expression that relates to the number of gray-scale values or colors that the image can display. For normal text processing, there are only 2 levels of gray-scale values available, black or white. This can be accommodated by one bit, either a 1 or 0, to represent either of the 2 extremes. More common gray-scale depths are 4 bits, or half a byte, 8 bits, or even 16 bits. Eight bits yields 256 possible colors, and 16 bits provides possibilities of up to 16 million [7].

9.2.1.3   Compression Ratio

There are several compression algorithms that may be applied to video streams. These will be discussed in more detail elsewhere in this book; however, of these different types, there are 2 different categories, namely, noisy or noiseless. Noiseless algorithms are those that maintain the quality of the original image or set of frames.

9.2.1.4   Noise Level

Noise level is the ratio of aberrant pixels to normal pixels. Aberrant pixels are those that have significantly differing values relative to their neighbors. Noise is unwanted variations in the presentation of information such that these unwanted variations cause unusual changes in the image relative to those values of the nearest neighbors of that pixel.

9.2.2   Audio

Vocalizations are the most important means of modern communication, because of the personal nature of such a medium. Even images have trouble conveying the emotions of the participants as compared to the emotions conveyed by audio. Ordinarily, audio can be provided by the network either as a pass-through option or as part of a multimedia activity utilizing digital formats. As a pass-through option, it can be digitized or analog in format. Obviously, digitized audio can be more easily manipulated than the analog form, and therefore is preferred when used in conjunction with other media. Audio information is ordinarily digitized in the system according to an algorithm specifying whether it came from some other recorded medium or directly as primary input from a microphone. This digitization process is referred to as a primitive function.

9.2.2.1   Musical Instrument Digital Interface (MIDI)

Beyond this process, audio is manipulated and packaged according to higher-level programs. One of these higher-level applications is known as MIDI, which will be discussed in more detail below. The subject of MIDI is important to the discussion of network management, because MIDI is a form of local area network (LAN) that manages the distribution and routing of audio or, more precisely, music [7].

One of the most common audio media architectures is referred to as MIDI, for musical instrument digital interface. This is a standard that is accepted worldwide as the definitive digital audio interconnection and networking protocol by virtually all makers of electronic musical equipment. A graphical representation of MIDI is given in Figure 9-4.

    MIDI, however, is much different from just the digitization of audio information or the networking of nodes. It represents an architecture and command structure that can identify different devices, command different nodes to operate in unison, and provide a code for the replication of different sounds rather than the actual sounds themselves. As shown in Figure 9-4, a primary or command unit can command another unit to perform any of the following tasks:

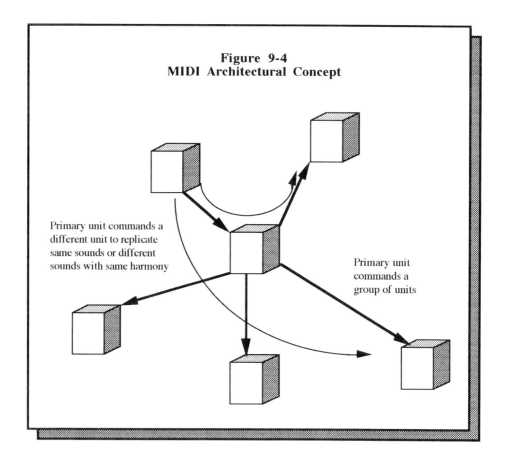

Figure 9-4
MIDI Architectural Concept

Primary unit commands a different unit to replicate same sounds or different sounds with same harmony

Primary unit commands a group of units

(1)   Replicate the sounds initiated by the first unit

(2)   Generate sounds different from those initiated by the first unit, but in harmony with them

(3)   Command 2 or more units to generate the same sounds, or sounds in harmony with those of the first unit

9.2.2.2   Audio Quality Measures

Audio quality is a matter of being able to faithfully collect and reproduce the source signal. The elements of a quality measurement relating to audio are as follows:

(1) $\left(\dfrac{\text{Sampling rate per second}}{44000}\right)$ -

        The greater the sampling rate, the better will be the quality of the reproduced version of the original signal capture. The generally accepted maximum sampling rate at present is 44,000 samples per second. Thus, any sampling rate less than this would be of lower quality.

(2) $\left(\dfrac{\text{Sampling depth}}{16\text{ bits}}\right)$ -

        Likewise, the degree of granularity of the sample is as important as the rate, because each sample has a range determined by the frequency components of the overall signal that make up each sample period. If this value cannot be accurately captured, the determination of the frequencies involved will suffer, and thus the quality.

(3) $\left(\dfrac{\text{Number of devices per net}}{1}\right)$ -

        If the various components of the system are part of a network in which any of the constituents can control other components according to some prearranged standard, such as MIDI, the quality of the overall presentation will be maximized due to the harmony and synchrony realized. The test for this quality is the presence of such a network and the number of devices on the network.

(4) $\left(\dfrac{\text{Response of pickup device}}{\text{Response of output device}}\right)$ -

        The quality of the audio delivery is also affected by the response of the pickup device as compared to the output device.

The overall quality measure can be expressed as follows:

$$\text{Audio quality} = \frac{\left(\dfrac{\text{Sampling rate per second}}{44000}\right) \left(\dfrac{\text{Sampling depth}}{16 \text{ bits}}\right)}{\left(\dfrac{\text{Number of devices per net}}{1}\right) \left(\dfrac{\text{Response of pickup device}}{\text{Response of output device}}\right)}$$

where

Sampling rate per second	=	rapidity of samples taken during any sampling period
Sampling depth	=	degree of granularity for each sample
Number of devices per net	=	number of audio devices attached to and controlled by the network
Response of pickup device	=	Frequency range over which the pickup device is effective. If the signal being sampled has a wider bandwidth than the input microphone, some of the signal frequencies will be lost, and therefore the quality will suffer because of such losses.
Response of output device	=	Frequency range of the collected signal is subjected to an output device that has its own response range. This range may limit the final audio delivery.

9.2.3   Text Information

Human communication through computer systems would be impossible if it were not for the computer's ability to process text characters [3]. This facility is aided by the internal coding of text, its ability to be compressed, and its ease of display through a combination of hardware- and software-based character generators. The following is a brief look at this essential medium.

9.2.3.1   Background

Text is the oldest and still most used medium for information exchange. Text was the first communications medium represented in computer communications. This first example of computer text manipulation involved the breaking and decoding of encrypted German communications traffic during World War II. After war broke out in 1939, the British knew that, in order to maximize their meager resources, they would have to anticipate, correctly if possible, German moves. The only way this could be done was for the British to read the Nazis' mail. The electronic version of this paradigm was for the British to snatch German radio communications out of the ether, and then to read the messages [3].

These messages were encrypted, and therefore had to be deciphered in order to be read. The deciphering process was a difficult one and required a considerable amount of manipulation, transformation, and substitution, all tasks easily handled by a computer. The British developed a device that could handle these tasks quite easily, and that later was recognized as being the first computer. The volumes of messages and computation required created a chaotic situation which necessitated a high-speed text processing system with large capacity. The system developed by the British had coded representations of the alphanumeric characters, string manipulation capabilities, concatenation, input and output interfaces, and, most important, algorithmic control over the various processing tasks.

Text processing itself operates upon a system of codes that represent the letters of the alphabet, numbers, and special characters, such as @, *, and #. Over the years several coding systems have been developed, most of which have been discarded in favor of EBCDIC and ASCII. ASCII is, by far, the most ubiquitous of the group. This coding system is flexible in that it can also be compressed for storage reasons, and it lends itself to displayable icons.

9.2.3.2   Text Quality Measures

The object of text transfers is to convey symbols from one point to another. There are numerous methods that have been developed to convey these symbols representing language. Additionally, special symbols used for formatting and presentation control, which are not displayed, also have a tremendous impact upon the definition of text quality.

Text quality measures vary depending upon the size of the presentation medium, such as the hard copy surface or the video display terminal being used. Additionally, the video terminal may have a varying degree of pixel spacing that either enhances the text presentation or limits it as the case may be. The types of questions that the quality measure may want to answer are:

(1)   How does the text appear on the visual presentation display device; that is, can it present the information in book format?

(2)   What is the sizing of the pixels for the video display of the text information, such that the presentation appears naturally presented for the viewer?

The answers to these questions allow the viewer to judge the quality of the presentation. This quality assessment can be based upon the following measures.

(1) $\left(\dfrac{\text{Pixels in horizontal scan/pixels per character}}{80} \right)$ -

The number of pixels in the horizontal scan, i.e., display dots per line, dictates the number of characters that can be displayed for each line. Special characters used for formatting are not considered in this equation, because they are hidden from the displayed format. If the displayable line is not capable of at least 80 characters per line, the display lacks sufficient capability for quality displays. In this regard, the ratio shown above has to have a value of 1 or more in order to indicate a quality display. The parameters are:

Pixels in horizontal scan   This value represents the number of pixels that occupy the horizontal component of the display. The display usually consists of 480 pixels across by 360 pixels down the display, conforming to the accepted ratio of 4:3, known as the display ratio.

Pixels per character   Each character in a font set is composed of an arrangement of pixels that convey the image of the character. This arrangement takes up a set number of horizontal and vertical pixels to define the character image of interest by displaying the font with a black pixel and the font background with a white pixel. The value of 80 was chosen in order to provide some degree of estimate for the proper breadth of character display.

(2)   The answer to the second question depends upon the size of the display screen, the size of each display pixel, and the spacing between pixels. The highest-quality displays have the smallest possible pixel sizes, minimum distance between each pixel, and maximum number of pixels per display screen size. There is no exact value for this measure, but each display of a lesser-quality display system will yield a less desirable display than the best one. Manufacturers today typically specify their high-end display products VGA, with pixel spacing of about 0.423 mm between display dots. This value is derived by multiplying the horizontal span of the display area, 8 in., which is equivalent to 203.2 mm. This value is divided by the total number of pixels per scan line, 480 pixels, the result being 0.423-mm interpixel spacing. For lower density, or fewer pixels per scan line, such as 360, the spacing is 56.4 mm. The point is that as the interpixel spacing increases, the coarser the display appears to the viewer. This effect makes graphics images gritty and distorts color applications.

The typical display will have a pixel matrix configuration that resembles that presented in Figure 9-5. The 2 major aspects of the text and graphic display are presented in this figure. First, graphic figures, which usually consist of at least some line segments, are subject to a certain distortion known as aliasing. This condition results whenever a desired line segment cannot be represented by a contiguous set of pixels in the direction of the intended line segment. Instead, some zigzags must be made by 2 or more smaller line segments that are connected by yet another small line that cuts across the zigzag lines to connect them. The second aspect of pixel display technology that is worthy of discussion is the presentation of text characters. These are portrayed as black pixels arranged in the pixel matrix so as to resemble the characters and special characters of the particular font style replicated and employed. Even such features as serifs are incorporated into the style and design of the particular font being used.

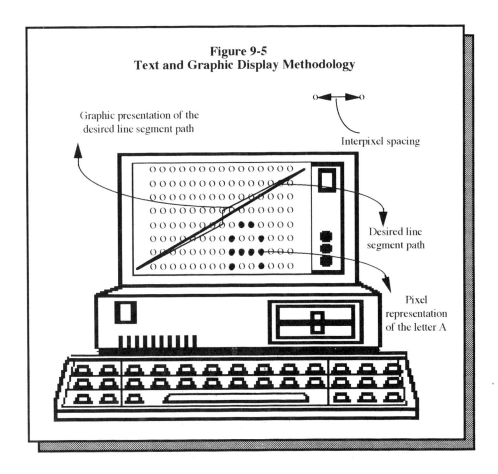

Figure 9-5
Text and Graphic Display Methodology

There are 2 ways that text can be portrayed on the display screen, one being to generate each text font by manually turning on and off each pixel required to portray the character desired. The other method is to use a character generator that employs font descriptions, stored in software, that are drawn upon when required to feed a character generator that in turn feeds the display. In the early days of computer systems, character sets were stored in the hardware environment, known as a firmware location, and from which the character generator drew when necessary for the display of individual characters. Additional software features in the system can also perform anti-aliasing so that the zigzag effect is minimized, and when the fonts are finally printed on a sophisticated ink jet or laser printer, the smoothest possible presentation of the fonts is created.

9.3   Predictive Methods

Predictive types of control techniques may be used to predict and regulate multimedia process control systems [8]. Typical examples of how predictive techniques may be used include predicting information transfer times between interactive points of communication, errors that may occur within some information interaction, and possibly certain aspects of image processing applications.

9.3.1   Polynomial Regression

The use of predictive methods can be of value when trying to predict the operability of a multimedia network [8]. Such techniques are valuable after the network has been up and running for some period of time.

The notion of future prediction is, to some extent, based upon some type of trend obtained from previous data in the series. This trend can usually be reduced to some type of polynomial equation obtained from empirical data. Using the derived equation, the user can then predict an outcome based upon some input, chosen to reflect a point in the future.

In the absence of any solid data history, the predictive function can be, at least temporarily, generated using one of several functions, among which are the exponential and power functions. These will be explored later. First, however, we will explore how a generalized polynomial is derived. What we want to do first is to determine an equation of the form:

$$y' = \beta_0 + \beta_1 x + \beta_2 x^2 + \ldots + \beta_p x^p$$

If we apply a least-squares criterion to the solution of this equation to estimate the coefficients, $\beta_0, \beta_1, \beta_2$, to β_p, then we will obtain the equations:

$$\Sigma y = nb_0 + b_1 \Sigma x + ... + b_p \Sigma x^p$$
$$\Sigma xy = b_0 \Sigma x + b_1 \Sigma x^2 + ... + b_p \Sigma x^{p+1}$$

.
.
.

$$\Sigma x^p y = b_0 \Sigma x^p + b_1 \Sigma x^{p+1} + ... + b_p \Sigma x^{2p}$$

The solution of a set of such equations where there are 2 coefficient unknowns will require 2 equations, while 3 unknowns will require 3 equations, etc. The idea is to fit the lowest-degree polynomial that will adequately describe the data under examination.

9.3.2   Correlation Functions

Correlation is the process of defining a numerical relationship between 2 independent sets of events. For example, if we were to examine the relationship between crime in Chicago and banana imports into the United States, we would find a very strong correlation between these 2 event sets. It should be pointed out that correlation does not necessarily guarantee a cause-and-effect relationship, as illustrated in the example cited. It only shows a correspondence, or correlation, between the 2 data sets defined. As it turns out, the relationship between banana imports and crime in Chicago is the weather, or more correctly stated, the average daily temperature. Thus, correlation does not imply cause and effect, but rather a similarity of increase of rise and fall that supports a relationship between the 2 items.

The exact definition of the correlation function can be explained mathematically as indicated below, but in a narrative sense it can be explained as the ratio of (a) the differences between a chosen function and the actual points against which the function is compared and (b) the differences between the average of the actual points of interest and the points themselves. In equation form, the definition is:

$$r = \pm \sqrt[2]{1 - \frac{\Sigma(y - y')^2}{\Sigma(y - \mathbf{y})^2}}$$

where

r	=	the correlation function value
$\Sigma(y - y')^2$	=	the difference between the observed data and function derived from the data
$\Sigma(y - \mathbf{y})^2$	=	the difference between the observed data and the average of the observed data

9.3.3   Simulation

One of the primary methodologies for assessing multimedia system performance is queuing theory [8]. Simulation capabilities have been available in software packaging for many years and have been employed in a wide variety of applications from commercial operations analyses to sophisticated high-technology investigations of nuclear physics, space flight, expert systems, and the like. Its place in the arena of multimedia systems usage and management might best be applied through its detailed analyses and investigations of multimedia interactions among 2 or more participating users [2].

Simulation became an important tool for analyzing physical connection and routing capabilities for the phone companies in the early part of this century. A. K. Erlang was a mathematician and early pioneer in the study of congestion as it pertained to electrical, land-line communications [1]. His work added much to this body of knowledge as he used and extended queuing theory for the study of communications system design for the, then, young AT&T Company. The issue with queuing theory and other analysis tools is, however, not so much analyzing the system as it is understanding the sensitivities of the system.

In most design situations the problem is really one of how much capacity is needed to accommodate the traffic flow, in addition to what kind of capacity is needed. This question can be further broken down by answering the questions of how much the system is going to be tasked and how fast the system can respond. It turns out that these are the two major components of the queuing problem. Even though the emphasis here is to present queuing theory as a study method for communications, it is also a method for analyzing any other system where delays and service backlog can occur.

9.3.3.1   Single Servicing Points

There are two major distinctions to be made in this line of study, those involving one servicing point and those involving more than one servicing point. A servicing point is a place where a customer, whether it be human or data gram, is dealt with in some way, such as a business transaction or protocol conversion. For the case where there is only one servicing station, or server, there are only two parameters of concern, both of which we have already identified, namely, the rate of demand upon the service point, and the rate of servicing that demand.

Further distinctions can be made about the way(s) that the system demand can arrive and the servicing can occur. These distinctions range from the various types of distributions from which demand and servicing can be drawn, to confidence levels for those distributions. The confidence levels determine the maximum sizes of waiting lines, and therefore, the capacity required. For instance, referring to Figure 9-6, we see that a one-server or single-server queuing system has the two components previously mentioned, and requires the same constraints of discipline type and confidence levels in order to answer the questions at hand, namely, numbers of customers in the queue and time in the queue.

In Figure 9-6, there is a depiction of a waiting line and service station [1]. The waiting line is filled with customers, assembly-line components, or other items that describe use or reason for analysis. These customers have variations in their arrivals that give rise to an interarrival distribution for these items. The nature of this interarrival distribution, in turn, gives rise to the customer arrival distribution that is one of the key constraints of the system analysis. This arrival distribution can take any form for which a function has been described, such as exponential, Poisson, constant, normal, log-normal, gamma, etc. The characterization of the distribution is important to the proper analysis of the system, but in the absence of any such identifying type, a random or Poisson distribution is usually assumed. Depending upon the arrival distribution assigned to the problem at hand, the equations for solution may be slightly different.

Figure 9-6
Queueing Theory

Single-server queue

$\rho = y * T_s$

ρ = utilization of the single-server system

T_s = service time per customer

y = arrival rate of customers

General rule: If the value ρ is greater than 1, the number of servers required is the fractional number rounded up to the nearest whole, e.g., 2.1=3.

The servicing of the customers that arrive into the system is also subject to variations in completion rates. These variations may range from no variation, sometimes called constant service, to extremes that are similar to arrival rate distributions. Again, these variations are depicted in such distributions as gamma, exponential, Poisson, etc. The arrival and service distributions may differ from each other as well. For instance, an exponential arrival distribution may be coupled with a Poisson service discipline, and so on.

Regardless of the distributions describing the arrival and service performed on the customers coming into the system, each has a mean value assigned. These mean values are determined from the average of the variates that describe the distributions under consideration. For instance, the mean (L) of the arrival distribution may be 20 arrivals during some period of time, such as a second, or 20 arrivals/s. The service activity (T) may take 40 ms for each arrival, or 40 ms/arrival. The utilization (u) of the system would, therefore, be $r = L \times T$. Taking into account the differences in time, i.e., seconds versus milliseconds, the system utilization would be $r = 20$ arrivals/s $\times 0.04$ s/arrival $= 0.8$. This is a nondimensional quantity that represents a percentage. This percentage represents how busy the service facility is in serving the needs of the customers that flow through it. Thus, in this case, the service facility is 80 percent busy. If the service time were 35 ms/arrival, the service facility would be 70 percent busy, and so on.

A treatment of the entire subject is beyond the scope of this book, however, it is appropriate to say that much about the characterization of a queue can be understood, given some of the basics outlined here. A further treatment of the subject is provided in a very good reference document published by IBM and cited in the references at the end of this chapter [1]. Further, a sufficient appreciation of the subject can be had by considering the two key elements of the queue, namely, the service time and the arrival rate. In a network management environment the network's performance can be evaluated from about three viewpoints. The network is evaluated according to how well it speeds the handling of traffic, its design that constrains the loading at the front ends of the system or nodes, and how efficiently the system levels out and distributes the traffic flowing within the system at any point in time. Distributed systems were originally conceived, partially, in recognition of the need to control the arrival rate problem, and will be discussed next as multiple servicing stations.

9.3.3.2   Multiple Servicing Points

The other situation, in which multiple servicing points are in operation, is a little more difficult to analyze and solve. We shall discuss this situation briefly here, with a more rigorous discussion to be found in the references at the end of this chapter. Multiple-service-point queues sometimes appear to be single-server queues because the waiting lines are peculiar to individual servers, such as in the case of a supermarket checkout area. The test, however, is whether concurrent (parallel) facilities operate to provide the same service(s) to an arriving stream of customers. Multiple service queues may share common waiting lines or separate ones, and their size may be an indeterminate number of service facilities. The basic issue underlying the multiple service queue, as with the single-service queue, is to control the waiting time, and to keep the queue from growing without bounds by providing enough servers to manage the influx of customers.

There are several different scenarios by which multipoint servicing can be implemented. These are discussed further in the references at the end of this book. One basic situation, however, occurs when we assume an unlimited number of servicing facilities, or transmission lines, to accommodate our communication traffic needs. An equation that can be used to determine the number of lines necessary to support such a system (assuming a Poisson distribution for incoming traffic) is:

$$m(r) = lTs + Xr\ SQRT(\ lTs\ )$$

where

l	=	arrival rate of messages (as a Poisson distribution)
Ts	=	servicing , or holding time, for the communication line
Xr	=	required deviation from the mean of the Poisson distribution for inclusion into the distribution sample, i.e., confidence limit
SQRT	=	square-root function

The confidence limits of a Poisson distribution to be associated with this formula are provided in Table 9-1.

Table 9-1
Confidence Limits for a Poisson Distribution

Confidence Area to Be Encompassed	Standard Deviation to Include Area Defined
90.0	1.3
95.0	1.7
99.0	2.4
99.9	3.2
99.99	3.9
99.999	4.5
99.9999	5.0

An example of how an analysis using this technique might be performed follows. Suppose we are interested in finding the number of lines necessary to support a system link where the arrival rate of traffic is 20/s, the holding time is 0.4 s, and the required confidence limit must include 90 percent of the circumstances to be encountered. Working through the formula shows that the required number of lines to support these requirements is 11.68, or 12 rounded up to the nearest whole number. This formula, although directed at an infinite number of service points, does track somewhat closely with solutions to problems where there is a finite, or limited, number of service points. For this reason, it provides a good representation of communication lines necessary for various confidence-level requirements. The system designer may have several methods of determining connection needs, but this method provides the reader with a good feel for the process involved.

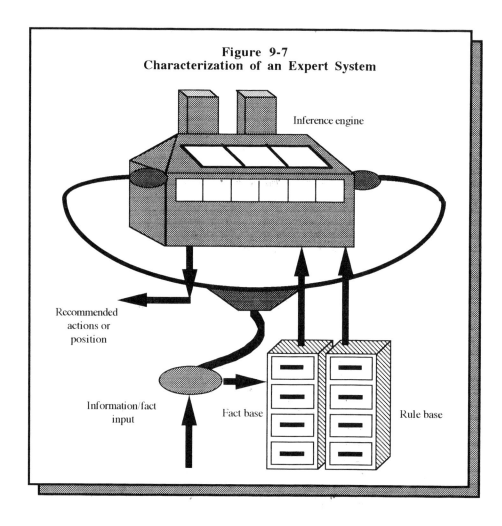

Figure 9-7
Characterization of an Expert System

Inference engine

Recommended
actions or
position

Information/fact
input

Fact base

Rule base

9.4   Intelligent Approaches

An intelligent system has certain characteristics that distinguish it from the traditional data gathering and analysis system. The basic approach to expert systems is depicted in Figure 9-7. In this figure the expertise is compared to a factory where engines grind away at the business of receiving data collected from the environment. As a departure from the traditional data collection systems, the engines use rules, rather than formulas, to make judgments about what the data mean and/or what conclusions are to be drawn from them. The figure shows an inference engine, or program shell, that acts upon rules previously established and receives input as to information and facts associated with the domain or area of expertise covered by the rules.

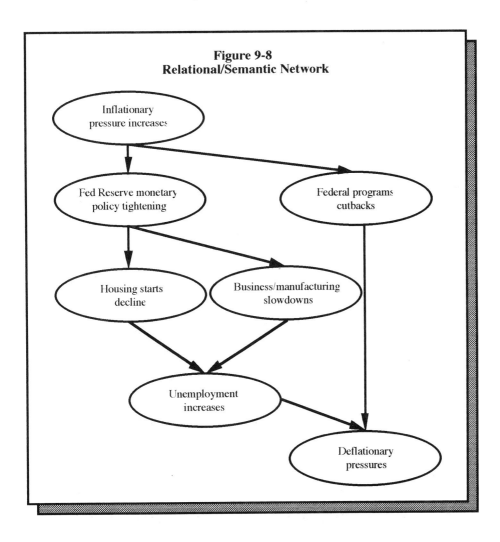

Figure 9-8
Relational/Semantic Network

The program shell assesses the facts in the context of the rules and offers recommendations or decisions as to actions to be taken. This section explores the various approaches to the design of expert program shells.

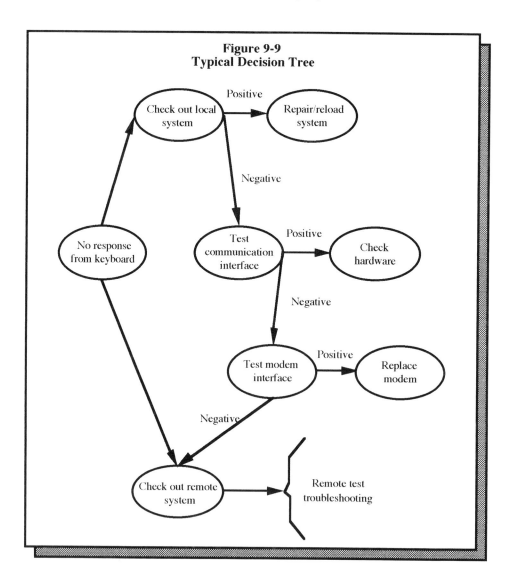

Figure 9-9
Typical Decision Tree

9.4.1   Data Structures

The key to the technology behind artificial intelligence (AI) is, of course, the computer. It is the closest thing to a working model of human memory, association, and problem solving, which has yet been created. The computer is limited in its ability to form associative references and to update its model of its world, but these limitations are likely to be overcome given the intensity of research and development in these areas. Network management uses of AI for the most part are to produce desired results within certain boundaries, for instance, the domains of fault detection and correction [3-5].

Knowledge is the stuff decisions are made of and can be represented in several ways. The major features of knowledge are form and content. Here we are talking about the structure of the knowledge represented, for example, as linked associations, as opposed to its current value. Knowledge has been represented in AI applications in several ways, which are summarized as follows [6]:

(1) Mathematics -

The traditional method of parameter representation is via symbolic quantities that are assigned values and value relationships relative to each other. Equations such as $y = x + 10$ express the necessary scaling of x to achieve the value of y that is essentially a comparison scheme.

(2) Relational or Semantic Net -

This is a structure that establishes the relational connections between objects along with a specification of those relationships. Examples of such semantic nets are operational interactions in a space vehicle, where vehicle maneuvering and payload operations must maintain relational connections in order to accomplish the mission. Another example is the organizational work flow in a company, where different departments must coordinate work flow in order to maintain a steady work output. Figure 9-8 shows an example of a relational/semantic network.

(3) Decision Tree -

This is a structure, similar to the semantic net, that provides relational associations but with the idea of leading to a "therefore" about the results. A decision tree provides a successive narrowing of possibilities leading to some conclusion or set of conclusions. A decision tree expresses a group of options and conditions available to the decision maker not usually made known to one using a semantic or relational net. For an example, see Figure 9-9.

(4) State Transition Graph -

This representation replicates operational transitions between
states of activity. It essentially provides a working view of some
process or machine. An example of such a process would be the
operation of a harvesting machine or the operation of a packet
assembler/disassembler organized in graphical form similar to a
semantic net. Each operational state is captured and allocated its
own identity.

(5) Frame -

This is a set of knowledge about objects and events that are linked
to a common situation. A frame about a network equipment item
might include performance ranges, fault indicators, and procedural
references. Figure 9-10 provides an example of a frame-based
knowledge representation.

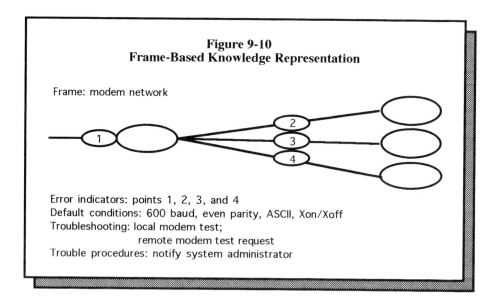

Figure 9-10
Frame-Based Knowledge Representation

Frame: modem network

Error indicators: points 1, 2, 3, and 4
Default conditions: 600 baud, even parity, ASCII, Xon/Xoff
Troubleshooting: local modem test;
                 remote modem test request
Trouble procedures: notify system administrator

(6) Logic -

The ability to infer a new set of facts from existing facts or to check
the validity of facts is referred to as logic. Such cases involve
inferred facts such as those given in the following syllogism: "All pit
bulls are vicious"; "this dog is a pit bull" therefore, the conclusion is
"this dog is vicious." The conclusion was derived as a result of
known or assumed propositions.

(7) Decision Space -

A graphical representation of relationships between parameters can be provided as a graph of independent versus dependent variables. Such a representation can also be used to portray more than two dimensions as well. Figure 9-11 shows a representation of the relationships between three parameters. A selection of parameter **A** yields a value for parameter **B**, which in turn dictates a value for parameter **C**. The association between parameters is made by intersection points on the functions where the variables involved intersect, as indicated in Figure 9-11.

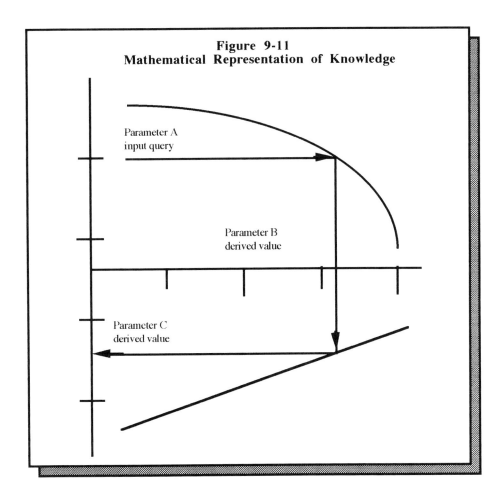

Figure 9-11
Mathematical Representation of Knowledge

Parameter A
input query

Parameter B
derived value

Parameter C
derived value

(8) Procedures -

A procedure is a step-by-step characterization of a process that provides the ability to transfer knowledge from one human or machine to another. Machines are very good at following procedures. A robotics device, for example, might have the following partial procedure installed: (1) raise manipulator arm; (2) move to X and Y coordinates specified in Table A; (3) open manipulator grasp; (4) lower to Z coordinate specified in Table A; (5) close grasp to texture W tension specified in Table B [6].

(9) Production Systems -

This involves the use of rules to achieve a degree of focus on the problem at hand. The use of if/then/else rules for computational suitability was first proposed by Emil Post while working at The City College of New York in 1943 [1]. This approach was later extended for use in studies of human intelligence as well. Examples of such rule sets are: If input switch set to X, then select output port B; or if input switch set to Q, then select output port K.

(10) Isomorphic Representations -

Similarities of structure between dissimilar objects, called isomorphisms, have been important to any number of investigations and subsequent discoveries. The investigation of differential transmissions of electrical signals has provided clues that have pointed to the same events occurring in the vertebrate brainstem. There neural signals from the same source are transmitted in two different paths to the interpretation centers of the brain [5]. They are thought to be used in the same way as differential electronic transmissions, where comparison of different signal paths can increase transmission reliability and eliminate noise carried by the system. Thus, the study of electronics may give insights as to similar mechanisms operating in biological tissue, and vice versa.

9.4.2   AI Tools

The acquisition and representation of knowledge has led to an array of specialized hardware and software tools that were developed specifically for AI. Such software tools included shells, or specialized interfaces, called by names such as KEE and ART. These software modules were based on LISP, the favorite AI programming language in the United States, whereas PROLOG has become the AI programming language of choice in Europe and the Far East. Since about 1987, however, the funding for commercial or government-sponsored programs that supported or furthered the development of AI and AI-based tools has decreased considerably.

The hardware supporting AI was developed and manufactured by companies specializing in this area. The basis for these hardware devices were specialized chips that could easily operate on LISP-based instruction sets by interpreting and executing these instructions as they occurred. This technology was, unfortunately, on the periphery of the mainstream open architecture moves that arose during the 1980s. This area of development has since given way to more general-purpose hardware and chips that are now being used for the integration of AI functions with more versatile platforms. Fortunately, the inclusion of AI functionality in with more generalized architectures has made it easier for AI-based approaches to be incorporated into mainstream product lines.

9.4.3   Knowledge Transfer and Reasoning

The particular circumstances of a situation and the requirement that knowledge be transferable may dictate the type of specific knowledge representation needed. Once knowledge is acquired and transferred to storage, it is said to have been learned. How do we impart knowledge to our systems? One type of learning is parameter learning. For network management, learning may come in the form of collected parameters that are compared. These collected parameters are facts that have been developed or received from sources external to the management system, i.e., within the network itself. In order to assess operational conditions, certain thresholds or ranges are provided to the system in the fact base for subsequent use with such collected data. The AI role in such implementations is in the judgment process used to assess the utility of these comparisons for later reactions to the system operation that may need to be taken. The question to be asked at these comparison junctures is, "So what?" Does an abnormal indicator mean that a problem exists that must be fixed, or does it mean a reconfiguration is necessary, or what?

Other types of learning include descriptive learning and concept learning [1]. Descriptive learning is the act of describing the problem or arriving at the solution from supplied signs and symbols. An example of descriptive learning is the ability to describe a chemical compound from certain characteristics of that compound. Descriptive learning used to be the exclusive domain of humans, as is concept learning. However, today there are rudimentary programs that, upon being fed a knowledge base of chemical characteristics and associated radicals and radical interactions, can derive a series of potential compounds that will satisfy the desired characteristics. Concept learning is the ability to generate new signs and symbols because the present set is inadequate for use. A classic example of concept learning was the creation of the concept for the DNA molecule, for which no model existed prior to that molecule's identification. Concept learning is many times too difficult even for a lot of humans.

Integral to the concept of learning is the form of reasoning used to do that learning. There are two major forms of reasoning: inductive and deductive. The *deductive reasoning* approach starts with a known set of facts about some aspect of reality and subsequently matches those facts with other incoming observations. The extent of matching indicates the level of confidence that one has in the resulting conclusions. With deductive reasoning, the basic truths are known, such as the generalized description of a duck. Collected observations are then used to see if they match what we already know about certain experiences, such as what a duck is supposed to look like. We all know that the fictional Sherlock Holmes was the master of deductive reasoning. He possessed a vast array of knowledge which, when challenged by observations, allowed him to match those observations with known facts.

In a network management environment, as in other situations in life, there are no absolutes, which means that observations may be incomplete or insufficient for a deterministic categorization of results. Thus, a problem in the system may not be entirely identifiable based upon the observations of the operation. Deductive reasoning is sometimes said to be reasoning from the general to the specific. A delay in message relay at some particular node may be indicative of such events as heavy loading, noisy communication circuits requiring an unusual number of retransmissions, poor protocol interfacing also requiring increased retransmissions, a soft, i.e., partial failure, somewhere in the node, and so on [3].

Inductive reasoning, on the other hand, is said to be reasoning from the specific to the general. Inductive reasoning is the type of reasoning used in research where there are no facts to draw upon about the problem at hand. Conclusions are derived from an increasing set of observations that tend to provide an increasing weight of evidence that will lead to one conclusion or, possibly, a set of limited conclusions. These conclusions are examined in the light of additional observations until enough confidence is accumulated to assure, to some level of probability, that the weight of observations describes a fact. For instance, observations about the operation of a piece of equipment are accumulated until some tentative conclusion is drawn about its structure or function. Once this is done, then subsequent observations can confirm a set of high-confidence conclusions. The bursting activity of a packet assembler, as an example, may led to the confirmation of certain conclusions about its operation.

Other terms that are important to the study of intelligence include forward chaining and backward chaining. Forward chaining utilizes the concept of deductive reasoning to establish a set of truths about some aspect of reality. This set is then used as the basis for comparison for incoming observations. For example, "If unit A and unit B do not respond to test message queries, then unit C is at fault." This is a rule that is predicated upon prior knowledge about the relationship between units A, B, and C. In this case, it is the application of deductive reasoning. Backward chaining starts with the conclusion that, for instance, unit C is inoperative. The questions then arise as to what symptoms would follow, such as the inability to communicate with units A and B, etc.

The current emphasis in AI is to use production systems to solve problems related to maintenance systems. A *production system* is the aggregation of the rule base, fact base and inference software that acts upon the incoming environmental information. Feature space, semantic nets, decision trees, etc., are related to the production system in that they provide the hierarchies and frameworks of knowledge representation used by these production systems. Using forward chaining, the if/then/else evaluators of information can utilize mathematics, semantics, or any other representations of knowledge to achieve deductive reasoning about the *domain*, or circumstances, for which the production system is valid.

The attributes of an expert production system that give it utility and value can also be its major detractors if the system is not properly developed and equipped. The knowledge database of rules necessary to activate and guide the user or system through the decision process can be quite large, and difficult to set up and organize. On the other hand, the knowledge base must also be nontrivial in content, otherwise its use provides no advantage. The rules themselves must be dealt with carefully when they are updated, because subtle changes can sometimes generate very drastic and undesirable consequences.

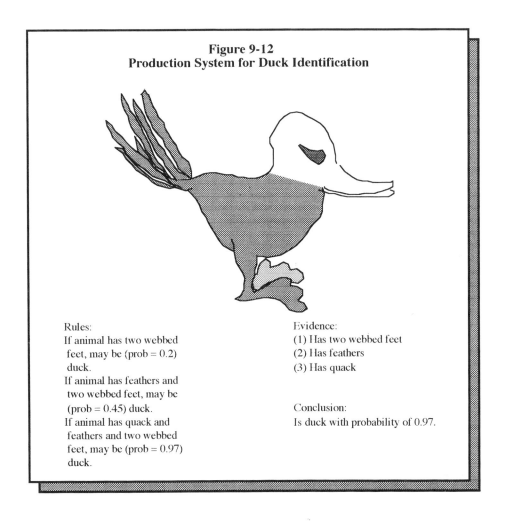

Figure 9-12
Production System for Duck Identification

Rules:
If animal has two webbed feet, may be (prob = 0.2) duck.
If animal has feathers and two webbed feet, may be (prob = 0.45) duck.
If animal has quack and feathers and two webbed feet, may be (prob = 0.97) duck.

Evidence:
(1) Has two webbed feet
(2) Has feathers
(3) Has quack

Conclusion:
Is duck with probability of 0.97.

A strong interface between the user and the system must be developed, because conversation between the system and the user occurs quite often. MYCIN was one of the first expert systems and was developed for the investigation of infectious bacterial diseases. The interaction between this production system and the user was intense and sometimes quite complex. In some production systems, frame-based representations have to be incorporated in order to help the user sort out organizing factors so that rules can be updated. An information frame is a self-contained set of facts that address a specific event within some organized context such as a modem network, as illustrated in Figure 9-10.

An interesting example of production system usage is as follows. Suppose you wanted to determine, on the basis of incoming facts, what kind of animal is being described. You have choices of decision making that are limited to horse, duck, and ostrich. The first fact is that the animal has two webbed feet. This rules out horse. The next fact is that the animal has feathers. This rules out neither duck nor ostrich. The third fact is that the animal quacks. This rules out ostrich and leaves the conclusion that the animal must be a duck, because it walks, looks, and talks like a duck. This scenario can be appreciated further by referring to Figure 9-12, where the rules, evidence, and conclusion are organized.

9.5   Design Control Approaches

Knowing what we really want helps us to focus on our objective. This is as true of network management systems acquisition as of anything else. If we can quantify what that objective is, we are in even better shape. The objective might be a transaction processing system, a file transfer system, a segmented multifunctional system, etc. Quantifying our objective(s) starts with understanding the parameters with which we are working. In network management, there are only so many parameters over which we have any control, and the important ones will be addressed below.

9.5.1   Reliability

Traditionally, reliability has been a measure of consistency, or consistent results. As a measurement parameter, it is calculated in several ways, such as by using the history of the equipment to make judgments, making use of parts counts to derive a complexity factor that can be converted into a reliability number, or evaluating risk factors in the technology that can also be translated into a reliability value [7]. Reliability implies the ability of a system or component to provide full-service capability between failures. For this reason reliability is somewhat analogous to mean time between failures (MTBF). Many times, reliability is the single most valuable measure for determining whether a certain system configuration is sufficient for operational needs or whether additional or substitute equipment is required. It can, more than almost anything else, affect purchasing decisions for a system.

If we have a system that has a certain reliability value, such as 75 percent, or 0.75, we know that the configuration of the represented equipment functionality will demonstrate its designed functional response about 75 percent of the time. Of course, there is no assurance that the response is appropriate; this would involve a different design parameter, namely, validity. If we are satisfied with the validity of the response but want a higher reliability value, something has to be done. We could look for another piece of equipment that meets the required reliability threshold, lower our expectations to fit the reliability promised, or use the current equipment in a redundant configuration.

The use of a redundant configuration also requires that additional equipment be employed to sense the failure in one piece of equipment so that the other may be committed to service. If we choose a redundant configuration where the failure of one system will trigger the use of another, what might we expect in terms of overall reliability? The answer can be reached by employing the following equation:

$$\textbf{TSR} = 1 - (1 - \textbf{SAR})^2$$

where

$$\textbf{TSR} = \text{total system reliability}$$
$$\textbf{SAR} = \text{stand-alone reliability}$$

$$\begin{aligned}\textbf{TSR} &= 1 - (1 - 0.75)^2 \\ &= 1 - 0.0625 \\ &= 0.9375\end{aligned}$$

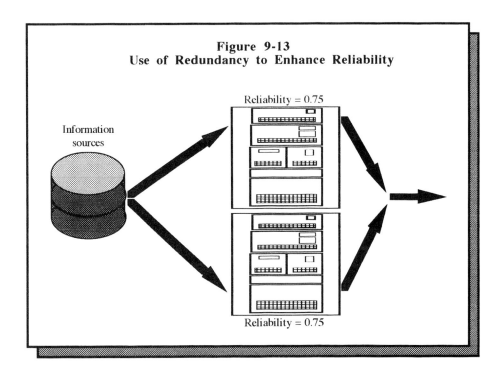

Figure 9-13
Use of Redundancy to Enhance Reliability

Thus, for the example cited above, if two units are to be used with a reliability of 0.75 each, the combined reliability will be 0.9375. This is the basis of fault-tolerant systems design. An illustration of such a problem is provided in Figure 9-13.

9.5.2   Availability

Availability is a measure of the system's ability to maintain its operational status, i.e., to be "available" for operations above and beyond its designed-in reliability [3]. The operational availability of a system also takes into account the system's scheduled and unscheduled maintenance times. If reliability is a measure of operational endurance between failures, then availability is a measure of total operational up-time that accounts for system reliability as well as its maintenance repair time. Quantitatively, availability can be expressed as follows:

$$\text{Availability} = \text{MTBF}/(\text{MTTR} + \text{MTBF})$$

where

MTBF	=	mean time between failures
MTTR	=	mean time to repair

Availability is always less than unity, because of the periodic necessity for maintenance or to accommodate unscheduled outages. It approaches unity as the reliability of the system approaches one or as the maintenance effort required for repair approaches zero. Today, more and more attention is being paid to the ease and facility for maintaining telecommunications equipment. Maintenance has been enhanced by use of several new concepts borrowed from the military, including modular replacement, referred to in government procurement circles as *line replaceable units* (LRUs), liberal use of high-reliability common transport paths, i.e., bus architectures, that interface all replacement units with each other, and the design of systems that diminish maintenance ambiguities to the maximum extent possible. The concept of availability is further portrayed in Figure 9-14 [4]. Both scheduled and unscheduled maintenance are incorporated into the system design and considerations for operations. Possibly, at some point, the maintenance activity can be automated, as alluded to with the robot that is shown in Figure 9-14, through robotics technology and/or artificial intelligence techniques.

9.5.3   Delay Times

In most applications today, speed is probably the most-mentioned design parameter. Its opposite is delay. Delay is expressed as a series of congestion points that tend to degrade the system's rate of response and operational efficiency. Generically, these points can be identified as processing delays, transport delays, queuing delays, and technology delays. A typical example of delays involved in a system operation is shown in Figure 9-15.

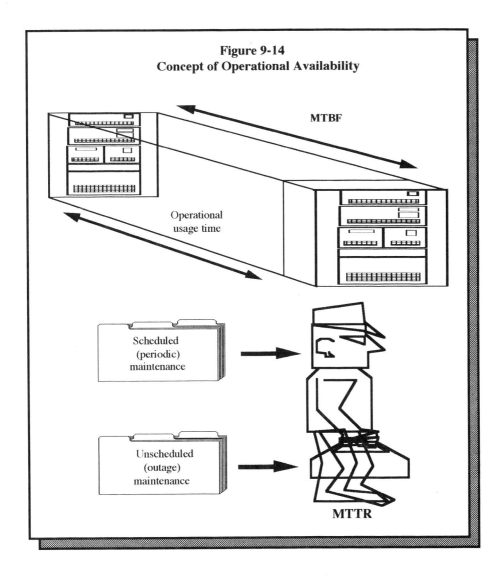

Figure 9-14
Concept of Operational Availability

Here we have an illustration of delays that exist in a typical transaction-oriented operation. The user creates a delay when entering a request, as the format and nature of the request is completed. Until the user presses the key to send the completed request to the computer, a delay is created while the data is entered. The request is queued up internally and eventually released for transmission, which also causes a delay in the system operation.

Next, the transport system creates one or more delays due to the transmission medium and protocols used to get the message from the origination point to the destination. At the destination, the request is again queued for processing. Depending upon the traffic to be processed, the request is serviced or processed, and if any confirmation or answer is due, that response is once again queued for the return trip back to the sender.

The delays experienced in a system are highly dependent upon the process involved, whether it be manual, completely automatic, or something in between. In some cases the origination of a request may be a human as used in this example, or it may be some automated feature of the system that automatically requests a task to be performed by some remote processor. Most of the time delays are a function of the data entry steps that are, however, becoming more and more automated as well. For example, bar code scanning of inventory items at a warehouse could serve as an automatic data entry request. There are many new technologies, techniques, and devices that are being developed to speed this part of the process and thus decrease the overall delay in the processing activities of the transaction chain.

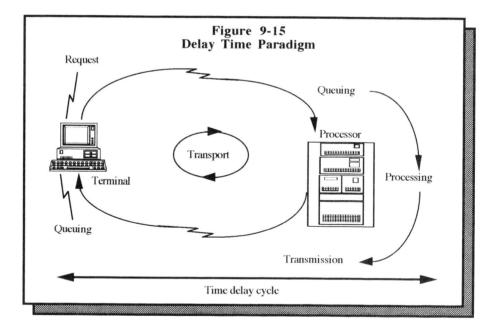

Figure 9-15
Delay Time Paradigm

9.5.4   Capacity

The capacity of a system to respond, react, or perform is one of the key aspects of that system's value, and hence one of the important system design parameters to be dealt with [1]. Capacity planning implies some form of statistical analysis in order to arrive at the required value(s). The use of this term also implies a reserve or safety margin for operations and rightly so. Any time capacity is mentioned, the assumption is that unforeseen circumstances are being planned for inclusion into the system. Unlike the parameters described above, capacity can be ascribed to many different aspects of the target system. There is the capacity of the communications transport system, the capacity of the message buffers that relay and process messages and/or requests in the system, spare capacity for additional system users, the capacity of the database system, etc.

Consider for a moment the ability of a system to deal with one type of capacity, namely, its input storage or buffer capability. Using some form of statistical analysis as indicated above, we can derive a value for this system's needs in this regard if we have some idea of the system's input rate. There are many different statistical techniques to use in such a situation, but the one that provides the random case for the clustering of arrivals, known as *bunching*, is the Poisson distribution.

The problem associated with this parameter might be expressed as follows: A system message input module is being designed that must accept all incoming traffic with a 99 percent confidence level that any incoming message will be received, not rejected because of a full buffer. If the mean arrival rate of incoming messages is 20/s assuming a Poisson arrival rate distribution, and a processing time of nominally 40 ms, also Poisson in nature, what is the buffer size required to accommodate this traffic?

The problem can be solved in a straightforward manner and is set up as follows: The loading on the system or its utilization (**r**) is a combination of the arrival rate and the processing time involved, i.e., $\mathbf{r} = \mathbf{l} \times \mathbf{Ts}$, where $\mathbf{l}$ = arrival rate and $\mathbf{Ts}$ = processing time. In this case $\mathbf{r}$ = 20 messages/s x .040 s/message = 0.80.

A formula may be used at this point that calculates the maximum number of messages (**p**) that may be expected during some percentage of the time, **r**. In this example, **r** is equal to 99 percent and is expressed as follows:

$$p(\mathbf{r}) = \left[\frac{\log(1 - \mathbf{r}/100)}{\log(\mathbf{r})} \right] - 1$$

$$p(99) = \left[\frac{\log(1 - 0.99)}{\log(0.8)} \right] - 1$$

$$p(99) = \left[\frac{(-2.)}{(-.096)} \right] - 1$$

$$p(99) = (20.83 - 1) = 19.83$$

Thus, for random arrivals and random service, represented by a Poisson distribution, requiring a 99 percent confidence level that the designed buffer will be able to accommodate incoming messages with nominal arrival rates of 20/s and processed at a rate of one message every 40 ms, a buffer size of approximately 20 storage locations is needed (rounding up from 19.83). The calculation may be slightly different for other distributions, but a random distribution represented by a Poisson distribution is very close to real-life experiences, and therefore this approach is appropriate for most cases.   This overall concept is shown in Figure 9-16.

9.5.5   Execution Speed

Another key parameter for evaluation when considering a network management system is the speed with which the system executes its tasks [6]. This speed is directly related to the system's inherent instruction execution speed but is also related to other features of the system, such as its architecture and use of algorithms in the delivered functionality.   The system's execution speed is achieved by either one of two methods, namely, by software program or firmware.

  Firmware is a combination of hardware and software where the executable object code is "burned" into hardware memory and is available for execution immediately, as opposed to having to be relocated into main memory from auxiliary memory for execution.   Firmware programs also execute faster because they are already loaded, decoded, and ready to execute in a form referred to as microcode.

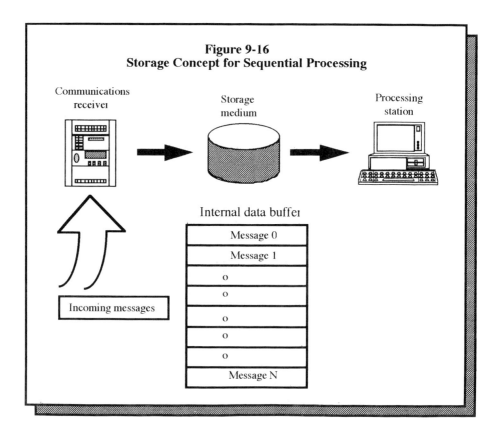

Figure 9-16
Storage Concept for Sequential Processing

A measure of the system's ability to execute instructions is referred to as MIPS (millions of instructions per second). This parameter is maximized by using firmware, streamlining the internal execution speed with concurrent use of more registers for mathematics and comparisons, streamlining the instruction set, or using a faster processor. A new concept in efficient instruction set usage is called RISC (reduced instruction set computer), which was developed because 80 percent of the computation in a representative program requires only about 20 percent of the normal complement of instructions. This has resulted in a redesigned CPU that has fewer instructions and increased capacity to handle many of the functions normally performed by using more registers within the processing unit itself instead of from memory. This has resulted in increasing computational speed.

Faster processors are also achieved through the use of either faster electronic technology and/or higher-density packaging of the transistors on the chip. Transistor technology has progressed to the point that it has achieved dramatic increases in speed as new strains of technology have been realized. The transition has run from PMOS (p-channel metal oxide semiconductor) technology that requires about 2 μs per instruction through NMOS (n-channel metal oxide semiconductor) that requires about 1 μs per instruction, to CMOS (complementary metal oxide semiconductor) that is somewhere in between the two, and bipolar technologies that can execute at between 70 and 100 ns per instruction.

Figure 9-17
Densities of Various Transistor Packing Technologies

Technology packaging	Density transistors/chip	
SSI	1	- 10
MSI	10	- 500
LSI	500	- 20K
VLSI	20K	- 300K
VHSIC	300K	- >500K

Note:
SSI	= small-scale integration
MSI	= medium-scale integration
LSI	= large-scale integration
VLSI	= very-large-scale integration
VHSIC	= very-high-scale integration

Notwithstanding all of this is the fact that the distance between transistors is also a major point of attack for technological pushes. All other things being equal, and given a constant speed, distance is the only other way to minimize time, since time equals distance divided by speed. As distance decreases, time decreases. Thus, the higher the packing density of a chip, the faster it can execute a series of instructions in order to accomplish a task. The data in Figure 9-17 present the relative speed advantage inherent in these newer technologies.

Summary

The idea of quality control and monitoring is essential to the overall problem of multimedia network control [8]. The various media involved in the multimedia experience must be carefully evaluated to assess the quality measurements needed to keep the user and administrators of the network apprised of the status of how these media pieces are functioning within the context of the applications. Also, several techniques are discussed in this chapter that relate the network control to necessary control measures, such techniques including expert systems, operational measures, simulation, etc. Several predictive methods are discussed that provide the user with methods of planning the management of the network more properly.

References

(1) *Analysis of Some Queuing Models in Realtime Systems*, IBM, White Plains, NY, 1969.

(2) Goyal, S. K., and R. W. Worrest, "Expert System Applications to Network Management," in *Expert System Applications to Telecommunications*, J. Liebowitz (ed.), John Wiley, New York, 1988, p. 1.

(3) Harris, C. J., and I. White (eds.), *Advances in Command, Control and Communication Systems*, Peter Peregrinus, London, 1987.

(4) Henning, W., "Bus Systems," *Sensors and Actuators A*, 25-27 (1991), pp. 109-113.

(5) Hoffman, M., "Technology Profile Neural Networks," *Techmonitoring*, SRI International, July 1991.

(6) Lemmon, A., *Marvel - A Knowledge-Based Planning System*, GTE Laboratories, Internal Report, 1986.

(7) McGrew, P. C., *On-Line Text Management: Hypertext and Other Techniques*, Intertext Publications, McGraw-Hill, New York, 1989, pp. 40-50.

(8) Yager, T., *The Multimedia Production Handbook for the PC, Macintosh, and Amiga*, Academic Press, New York, 1993.

Chapter

10

Multimedia Network Management Architectures

Chapter Highlights:

The network management of multimedia networks is very much an issue of applications management. Network management systems of the past have been designed and developed employing special-purpose hardware, without much regard to the content. Network management in the past has been a matter of monitoring and diagnosing equipment and transmission path failures. The content was affected only when such problems occurred. Multimedia network management is affected by additional criteria, such as the technologies discussed in the previous chapters. For this reason, multimedia networks and management architectures must be analyzed and synthesized on the basis of the outputs resulting from the operative features in the applications programs. Multimedia communications are more an issue of qualitative results than quantitative ones. Thus, the margins between acceptable and unacceptable outputs become smeared.

10.1 Architectural Concepts

10.1.1   Open Systems Interconnect (OSI)

There has been recognition for several years that communications networks should be vertically segmented and categorized for better identification and differentiation of unique functions or physical attributes. This segmentation has led to the creation of the International Standards Organization (ISO) model for networked communications, initially published in 1984. Figure 10-1 shows that this model, which is known as the open systems interconnect (OSI) model, has segmented the entire range of communications interchange into 7 layers.

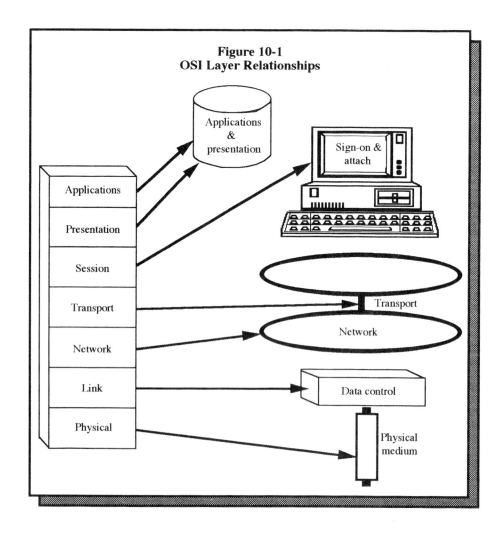

Figure 10-1
OSI Layer Relationships

Previous to this time IBM, DEC, and others had devised their own versions of how networked communications should be organized, but these schemes have given way to the more standardized one approved by the international community.

Figure 10-1 shows the relationships of the layers to each other and to the physical or functional world in which they interact. The ISO standard for the OSI model describes the functionality necessary for communications between layers and across layers to peers or equals at other nodes and termination points. This model has become one of the major foci for network management systems architectures and product development since the OSI model was adopted.

In the years since the adoption of the OSI standard for network management, many different protocols have been developed and refined to conform to one or more of the various layers of the 7-layer model. One protocol standard that is part of the OSI 7-layer standard is known as the Common Management Interface Protocol (CMIP), and is designed to encompass all of the OSI layers and interfaces as they have been developed and have evolved for each layer [19]. It is the OSI network management protocol and addresses the various monitoring and control tasks that are part of the model.

Because of the lead time required to realize the promises of the CMIP of the OSI model, some segments of the telecommunications industry in the U.S. felt that it was necessary to develop interim network management protocols. This sense of urgency was underscored by ongoing Department of Defense (DoD) Requests for Proposal (RFPs) that required network management facilities for the DoD's new or upgraded worldwide communications systems.

One of the other protocols that were designed and developed to support network management was the Simplified Network Management Protocol (SNMP), which became an often-proposed solution to the ongoing DoD requirements [16]. It is a protocol that is currently in place, has vendor and user backing, and is implemented in many product offerings by several vendors. In short, it is ubiquitously available to provide network management capabilities, as an alternative to the full set of network management standards promised by the OSI model. The SNMP protocol addresses at least 2 of the OSI layers and utilizes standards originally specified by the DoD.

Whether or not SNMP was originally envisioned to be an interim step on the path to full CMIP implementation in all networks is irrelevant at this point. SNMP now has industry and user support, partly because it has a certain "no nonsense" atmosphere to it that makes it attractive as a potentially permanent fixture in the network management world. In addition, it is based upon standard and layer definitions similar to the OSI model. For this reason, it may eventually become an integrated part of the OSI framework. Its major drawbacks are that it does not have universal endorsement, and it addresses only 2 or 3 of the layers of the model.

The Open Systems Interconnect (OSI) model segments the communications process into a series of layers that have functional characteristics, and for which standards of operation can be written [20]. These layers range in nature from the tangible or physical at level 1 to the creative or applications-oriented at level 7. The CMIP utilizes the OSI definitions for each layer to exercise network management tasks in each of these segments. There are several standards that have been initiated or are in process to support the various OSI layers. Figure 10-1 shows the relationship between the defined OSI layers and the physical and functional structure of the system. For network management purposes, a network management system is hard to define until some functional activity is present, at least at the physical or lowest level. In the meantime, standards have been produced that support functionality at each level, and upon which network management development can be created. The standards work will support interface specifications between levels that can stimulate protocol interchange between equivalent layers.

The definitions of the 7 layers of the OSI model are provided later in this chapter but are outlined briefly here. The 1st layer of the model provides the physical and electrical connectivity necessary to interface nodes or terminals to each other. The connection need not be entirely physical, as, for example, with space-to-ground communications that rely upon electromagnetic energy, amplified and transmitted through the ether to satisfy the connectivity requirement.

The key ingredient in this case is the need for "electrically powered transmission" between points of communication. The 2nd through 5th layers, described below, provide the network support necessary to get information from one point to another. Layer 6 is essentially the language base for interpretation of the information at either end of the communication. The information generated in the 7th layer is encased in data structures that are generated by applications programs, also resident in the 7th layer.

10.1.2   The ISO Environment

The ISO model, as mentioned previously, is an international standard for telecommunications and is a vehicle for the definition of a network management system. Figure 10-2 shows the interface between CMIP and its associated ISO layer interfaces. It is based upon the concept of layering such that as information is passed from a higher layer to some layer below, the functions become less complex and variability of use becomes more restricted. The idea behind the layers is to differentiate them on the basis of technology or process. Like any other standard that seeks to pigeonhole functionality, it has had both positive and negative influences. The buyers of systems have tended to think of equipment as fitting into one layer mold or another. It could be argued that this has tended to stifle creative development in some cases. On the other hand, layered definition has led to a taxonomy of thinking and understanding for most users and buyers that has benefited the use of equipment operating at each of these levels. Each of the key features of the ISO and CMIP interaction is discussed in more detail below.

10.1.3   Common Management Information Protocol

The network management functionality performed in the ISO environment is supported by the Common Management Information Protocol and its companion Common Management Information Service (CMIS) [19]. It will eventually interface with all 7 functional layers and provide the translation of status and commands necessary to perform network management at all the layers. CMIP operates through a set of layer management entities (LMEs) that provide specific interface to each layer. These LMEs, in turn, interface with a system management application entity (SMAE) at each network element to code and decode the CMIP information passing between network elements [9].

10.1.4   Layer Management Entities

The LMEs receive information from each layer relative to the status of activity occurring at that level. Each layer is a configuration and set of components that perform some definable function, such as packet transport. As a rule, the smaller and more contained the subject, the better is the control over that activity. Using this philosophy, one can imagine that a constrained functional activity at one of the layers can yield very specific information about that activity. This is part of the reasoning behind the idea of layered management.

Each LME concentrates on the functioning at its level of responsibility; thus, any indications of performance and/or malfunctioning tend to be very specific about the entity at that layer. Unfortunately, the risk is that the whole story about system performance and/or malfunctioning may not be told at any specific layer, because of the complexity of the system. Information at any layer may be indicative of problems in another layer as well, or exclusively in some other layer, thus, causing some degree of ambiguity.

10.1.5   System Management Application Entity (SMAE)

The status at each layer must be sifted, assimilated, and integrated with each network management function performed at that layer along with functions performed in the other layers in order for the total status picture of the system to become clear. This integration and interface function is performed by the system management application entity. Thus, the layer status is accumulated by the LMEs and passed to the SMAE at each segment in the system for analysis. Additionally, the SMAE provides interface between LMEs at one node and LMEs at another node of the network through the Common Management Information Protocol.

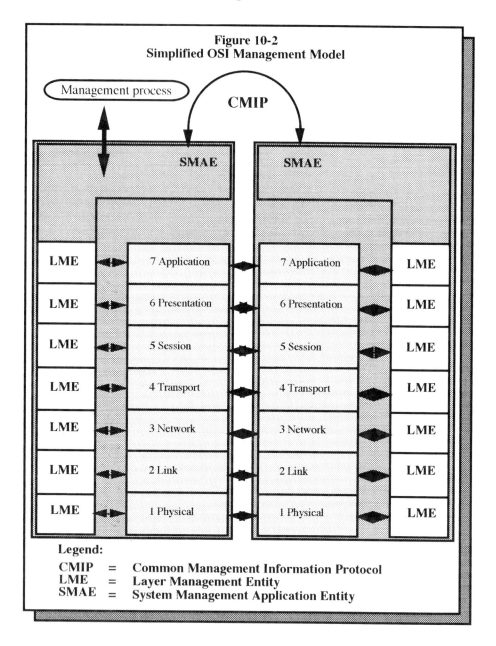

Figure 10-2
Simplified OSI Management Model

Legend:

CMIP = Common Management Information Protocol
LME   = Layer Management Entity
SMAE = System Management Application Entity

10.1.6 Development Approaches to Multimedia Networks

The approaches to network management have several possible routes of development. These various choices are illustrated in Figure 10-3. This figure is a chart of approaches based upon whether the system is to be integrated, decentralized, or specially developed for the application network in use. The costs, development times, and complexities involved all vary as a function of the particular approach used. A brief discussion of the evolution of operational networks will aid in better understanding the problems of network monitoring.

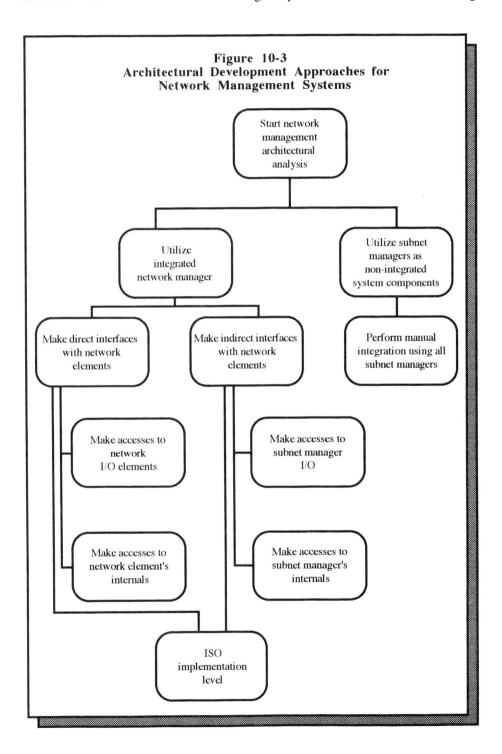

Figure 10-3
Architectural Development Approaches for
Network Management Systems

The next few figures will illustrate various networking configurations, each of which will be briefly elaborated upon along with the diagrams.

Point-to-Point -

The first part of Figure 10-4 shows point-to-point communications where the communications between remote facilities are direct, without intervening nodes, or interface points, and operating on one level of protocol interchange. This approach to networking is very limited, since communications cannot include more than 2 points of interaction. For multimedia, this approach is the best possible situation. Multimedia has evolved beyond this simple connection scheme to one where all users must have acces to all other users. Thus, a network management system based upon such a basic format is unrealistic.

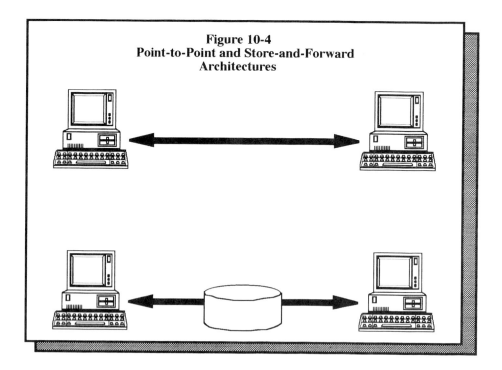

Figure 10-4
Point-to-Point and Store-and-Forward
Architectures

Store-and-Forward -

Certain forms of communications rely upon the transmission of information to an intervening point for later distribution of the information to the destination point. This form of communication may operate on different levels of protocol interaction, but the transmission system involved is still very simple. Error detection may be more difficult due to the added complexities of the system configuration. The second half of Figure 10-4 illustrates the store-and-forward paradigm.

Multinodal Network -

Multinodal networks are those equipment configurations that incorporate 3 or more terminals or stations that interact with one another. Such networks may interact via a protocol of digital messages, or through a switch that physically reroutes messages. The former is referred to as a packet switched system, and the latter is referred to as a circuit switched system. The exact nature of the multinodal network can be quite varied based upon the connection configurations possible. In addition, multinodal systems may have servers that might have several functions, among which could be to act as store and forward devices, switches, or repositories for applications programs, to name a few. Figure 10-5 shows the possible scenarios associated with this approach. On the one hand, the multinodal network may consist of virtual circuits connecting each of the terminals with each other. In such a case, the system is a packet switched network. If each of the terminals is connected to a switch that determines which of the terminals is being addressed, as indicated by the database symbol, the system is a circuit switched network.

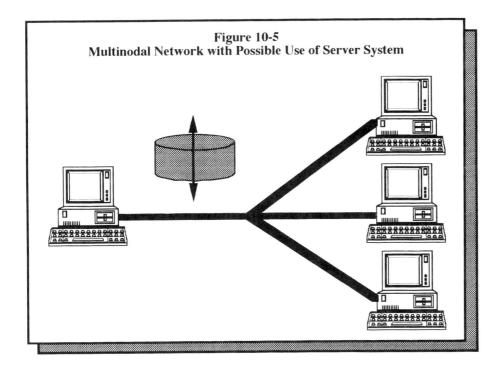

Figure 10-5
Multinodal Network with Possible Use of Server System

10.1.7   OSI Model Communications

In the OSI model, communication between peers or equivalent layers of the model is accomplished by a series of protocols that relays the desired messages through a series of interfaces and physical media to the equivalent layer in the functional unit being addressed.   The message path is down through the various layers of the model to the physical level, across the physical link between communicating nodes, and back up through the OSI layers to the layer equivalent to that initiating the communication.   For example, presentation layer communication is accomplished through the exchange of display control, formatting, and encoding information between equivalent layer 6 nodes. Communicating network elements are programmed to accept and understand differing protocols, thereby allowing their interactions using the OSI standards.

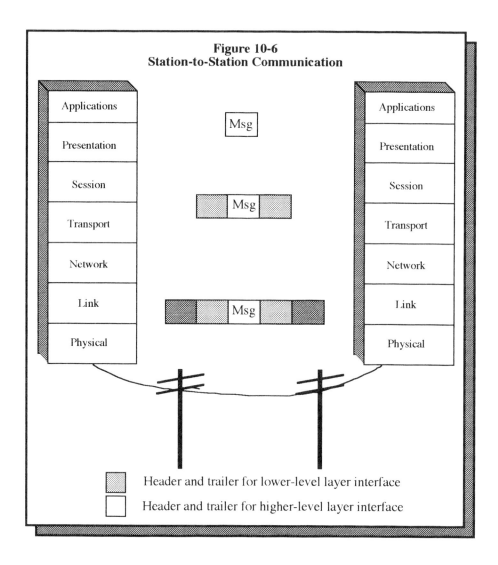

Figure 10-6
Station-to-Station Communication

This communication exchange between the 6th layers at two different nodes is gated by interfaces at the layer boundaries.

Additionally, each layer may have its own overhead information attached to it so as to identify that layer when it is passed through the layer of interest. Figure 10-6 illustrates this characteristic. Here just 2 sets of overhead information are used to illustrate the use of data packet headers and trailers to pass information. The message initiated at the 7th layer may have a header and trailer attached to the informational content of interest. As the message descends the OSI model layers, it assumes additional headers and trailer information prior to its release over the physical network. Once received at the destination, the message again ascends the OSI model layers, losing headers and trailers as appropriate at each layer until it arrives at the destination level, which in this case is the 7th layer.

The relay of information exchanges through the layers of each node and from one node to another is accomplished by appropriately packaging and stripping the contents of each transmission. This assures the integrity of information as it passes through each layer (much as each layer of skin in an onion protects each succeeding layer), and assures the proper regulation of that information. This process is the interface activity. The way in which information is packaged for transmission represents the mutual understanding or protocol implemented between nodes.

There are, generally, two types of protocols, one being electrical and the other being message oriented. The electrical-oriented protocol sets or raises and drops signal levels in order to signal interfacing equipment of the changing status of this unit. The message-oriented protocols set and communicate status or commands by passing coded status messages or commands back and forth. Each of the OSI layers has one or more protocols defined or in process of definition through the work of international standards bodies. This means that each standard may address a slightly different functionality, but each subscribes to the basic requirements of its targeted layer functionality. A summary of standards developed or in process of being developed for each of the layers is provided below.

Figure 10-6 illustrates how a protocol operates. A message works its way down one chain of the communications layer stack and back up the other side to the ultimate destination. This process requires administrative control at each layer. At each level of administrative control, information is attached to a message or message part, and the message length changes. As the message travels back up the other side of the message protocol stack, this administrative control is stripped until the message reaches its destination. Peer-to-peer communication is accomplished indirectly by transfer from one applications layer interface down through one set of layers and back up through the corresponding set of layers to another layer of equal rank at its final destination.

10.2 Multimedia Network Management

Multimedia networks may appear to be analogous to data networks, complete with information scanners, display monitors, printers, and various input devices. Some additional hardware in the form of TV monitors, microphones, speakers, VCRs, camcorders, etc., may also be part of such networks, but the usage of the various modalities involved is the major issue associated with multimedia networks. Thus, referring to the 7 layers of the ISO network management model, the top 3 layers, namely, session, presentation, and application, are the main focus elements of multimedia network management.

10.2.1   Multimedia Network Management Requirements

The unique network management requirements for multimedia are defined below:

(1)   Continuous quality assessment and measurement for the duration of the media processing cycle.

(2)   Universal protocol detection and conversion to and from the storage medium.

(3)   Primary concern with the applications rather than the transport issues of the multimedia network.

(4)   The repositories for multimedia applications have a tendency to be fractured, with various storage protocols for different media types that have to be integrated for meaningful usage.

(5)   The type and variety of technologies associated with multimedia network management issues makes the total problem more complex than would be the case with data network management.

The control of any network, as well as multimedia networks, requires a sensory interface with the physical junctions, or points through which the various modalities traverse, or a functional interface with the information as it traverses a common transmission system. These approaches can be appreciated by referring to Figures 10-7 and 10-8. These diagrams illustrate the 2 major approaches to multimedia management. Where Figure 10-7 shows a total integration of network elements based upon direct access of the network manager to each of the elements, Figure 10-8 shows the access concept for management information based upon indirect access to the network information.

Network management information about individual network elements may be gathered from one of 2 major sources. First, it may be gathered from the communications ports of those units. Most telecommunications devices communicate their status via, either the port through which they carry message traffic, or through a separate port specifically designated for network management information. Second, it may be gathered from the subnet managers that manage a set of homogeneous components. This is usually accomplished by monitoring the printer ports of the workstations serving as subnet managers.

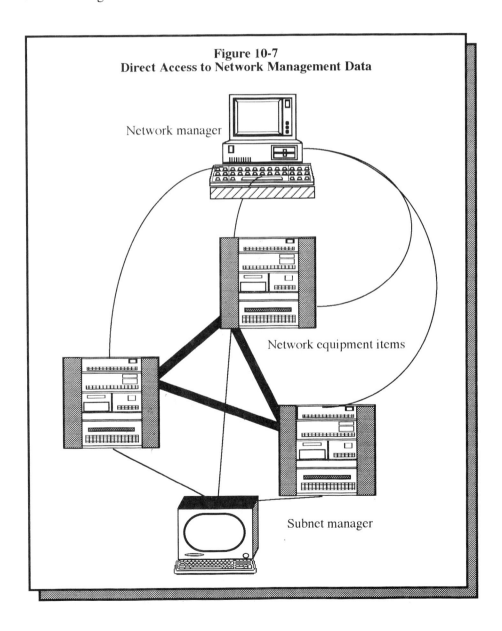

Figure 10-7
Direct Access to Network Management Data

Network manager

Network equipment items

Subnet manager

Periodically, the subnet managers will provide status and error messages to their printers, and such management messages can be provided to the overall network manager from the RS-232 ports that ordinarily send messages to attached printers.

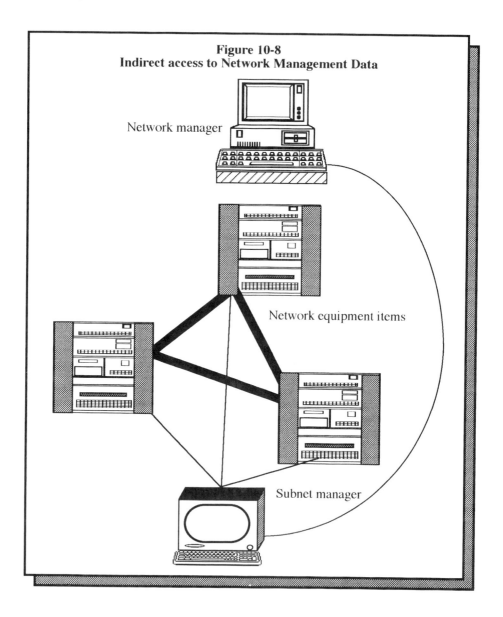

Figure 10-8
Indirect access to Network Management Data

Network manager

Network equipment items

Subnet manager

Integration of the multimedia signals at the local nodes may allow physical monitoring and control of the various modalities prior to their being integrated and multiplexed onto the internodal transmission links. During transport between nodes, the data are either packetized, or in some other way multiplexed so as to mask their identity. The management of this data during this transport period is handled by a database management scheme where tracking and auditing of the data is accomplished by decoding the data packets to identify the data, or by demultiplexing the frequency- or time-slotted data in order to read the data contents for purposes of identifying the contents and problems pertaining thereto. These 2 principles of multimedia network management point to a series of axioms that can be summarized as follows:

(1)   Status monitoring facilities of the constituent modal inputs and outputs should be located as close to the modal output as possible.

(2)   The modal inputs and outputs should carry source, and if possible, path information for purposes of isolating their sources and routes, and therefore their problems as soon as possible. In this way, at least physical components can be determined to be malfunctioning and thus replaced as quickly as possible.

(3)   The point of multiplexing, or packetizing, should be richly monitored with status messages.

10.2.2   Approaches to Multimedia Network Management

Multimedia systems are qualitative rather than quantitative, as data communications networks are. Data communications networks rely upon exacting standards of data, packet, and message sizes, along with error detection and correction codes, and the like. Multimedia networks, on the other hand, convey signs and symbols as the means of conveying meaning and impressions. These signs and symbols do not have to be exactly whole nor completely defined in order for the observing human to realize what these signs and symbols are trying to say and what their impact is. The overall defining criterion is similar to the exercise of vision as the means of sensing the environment around us.

The sensation of light rays reflecting off a set of objects back into our eyes, and therefore creating an interpretation in our brain of what is being observed, does not require all of the possible light rays to be so reflected. After all, the term *interpretation* itself means making a decision based on incomplete or seemingly disjoint information. There are basic approaches to multimedia network management as shown in Figure 10-9. In this figure, the physical and functional concepts are illustrated. Figure 10-10 shows how the integration of the various communications modalities are integrated.

The features shown have their own set of requirements based upon configurations and equipment assortments. The usual approach for mixed technologies is to monitor on the basis of LANs, nodes, or distribution/switching points.

Therefore, multimedia networks do not have to provide totally accurate nor complete information in order to perform their assigned tasks adequately. The measurement of their functions is performed at 2 different conceptual levels. One is the traditional monitoring and control of the physical and functional activities being conducted over the network and based upon its configuration. The other level of management is performed at the applications levels, which is where the interpretations occur. At this point, the voice, print, and video can be incomplete but yet still intelligible and still convey the message.

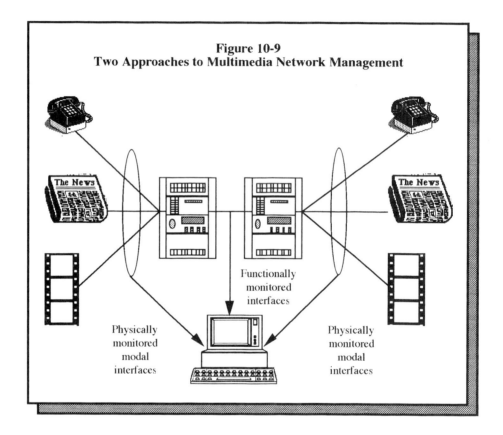

Figure 10-9
Two Approaches to Multimedia Network Management

10.2.3   Component Technologies of Multimedia Networks

While there are many significant economic trade-offs to be considered in implementing digital solutions to communications problems, such as the costs of the digital solutions, the flexibility and utility of digital solutions are unquestioned. Growing national and international trends are helping to assure the dominance of digital communications in the years to come. These trends include the conversion from analog to digital communications switches within the telecommunications systems of the world and the development of wider and wider bandwidth transport services that make increasing uses of digital applications possible.

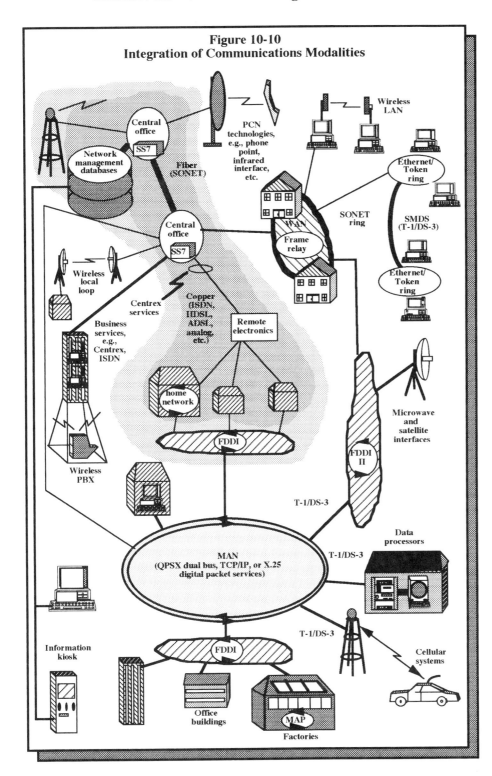

Figure 10-10
Integration of Communications Modalities

The question always arises, in designing or operating a network, as to the bandwidth capabilities of the system or its circuits. The trade-offs between data rates and distances achieved with differing rates of transmission were addressed above. Ideally, effective transmission distances should decrease as the transmission rates increase in an exponential format as shown in Figure 10-11. In reality, this is approximately true except that other parameters come into play, such as hysteresis, the nature of the transmitting and receiving equipment, the coding and modulation schemes used, the powering mechanisms, and the grounding systems involved. Limitations are usually codified in the various standards generated for transport capabilities, and distance restrictions, in the form of distance versus data rates, are also usually provided.

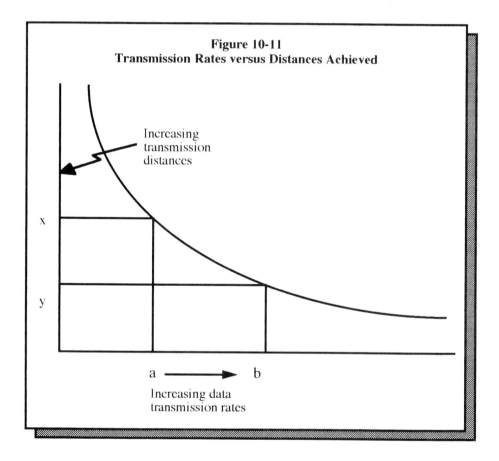

Figure 10-11
Transmission Rates versus Distances Achieved

Whether we like it or not, the basic line drivers and receivers in operation today, and for the foreseeable future, will likely conform to the RS-232 standard [16]. This standard provides for communication through telephone lines and between remote communications devices, such as modems. For this reason, RS-232 has been, and probably will continue to be, the principal physical means of accessing and interfacing networks. Newer standards, such as RS-449 for higher-rate data traffic, are being implemented, and even wider band standards are also being developed to bolster bandwidth versus distance requirements. The networking technologies and standards implemented to achieve the desired results also require that the various networks be capable of bridging to other networks so as to provide increasing access among many different people using the system under consideration. Indicated below are some of the major bridging technologies used in achieving this objective.

Any of the network architectures described above may be used in a local environment and may be part of larger systems that have periodic requirements to access other networks within the system. Certain functional separations may exist within the network because of specialized tasks performed within the overall framework, such as engineering or accounting, or because of some security separation requirements, such as finance or other less sensitive functions being performed within the organization. Periodically, however, the various members of the group may have a need to transmit to or request information from some part of the system that is not on its specific network. This is accomplished by one of several methods, such as bridging, routing, or use of gateways. The use of these interfacing devices may also increase ambiguity about the causes of malfunctions occurring within the system because of the increased complexity of the system configuration; this issue will be discussed later. Each of the principal network interface technologies is explained below.

(1) Bridges -

Bridges are essentially extension cords that physically connect one network with another. The hardware and software components that make up the bridge provide the ability for communications to be transferred from one network environment to another via a physical connection. Bridges are not concerned with protocol differences that may exist between the two networks involved. They are present to provide the interface, not logic or protocol conversion. The networks of interest may often be LANs requiring interface between similar network types, e.g., token ring to token ring or Ethernet to Ethernet, in a networked environment. The network bridge interface technique is probably the simplest method of facilitating conversations between geographically remote or logically separated entities within a multinetwork, system environment [16].

(2) Routers -

Routers are more sophisticated than bridges because they provide 3 different services. Not only do they provide connectivity between networks, they also provide routing functions from one to many networks or gateways, and at some point they will provide logic or protocol conversion as well so that dissimilar networks may talk to or through each other. Some companies already provide such protocol conversion facilities within their router products, but standards for this type of feature are still in the making. Routers provide logical as well as physical connections between networks of dissimilar type and, as such, become another important "moving" part in the total system that must be accounted for in terms of such issues as reliability. Protocol conversion is the task of decoding the contents of a formatted message and reformatting it for retransmission through another network that can recognize that format. The router's job, therefore, is one that is intrinsically more complicated than that of the bridge.

(3) Gateways -

Gateways are essentially permission-driven access points to any number of information channels that may be of interest to the user. "Permission-driven" here means that the access point to the various information channels is controlled and available only to those that have permission based upon passwords and the like. Examples abound in today's world such as financial networks, stock trading networks, personal shopping networks, and weather networks, all of which are accessible from one entry point. Gateways can also be used to access "read only" data sources that may be of interest to a requesting system user. "Read only" provides only access to certain information, without allowing the user to alter the data files from which the information is derived. For instance, someone wanting to determine facts about system usage for the purpose of developing a proposal for expanded capability may need to access information about system performance, costs, and current inventory. This information could be obtained via some sort of access point such as that provided in a gateway feature [16].

In a multimedia environment, all of these types of devices with their attendant functional capabilities will be available for multimedia communications. Because of the fact that packet communications appear to be the best trade-off for controlling multimedia delays, routers will have to minimize the effects of delays so as to provide a seamless connection between the media sender and the media receiver.

10.2.4   Protocols for Multimedia Network Management

Protocols are methods of interaction between functional units such as digital equipment and certain software-oriented environments. Protocols are desirable for many reasons, not all of which revolve around technical issues. Vendors are many times very supportive of standards for protocols in order to assure potential customers and ongoing clients that their products conform to industry standards, and have long-term utility and survivability. Industries are interested in standards as a means of establishing rules and keeping their activities organized. Users are interested in standards as a means of protecting their investments. It could be argued that all of this interest is motivated by money concerns, and while probably or at least partially true, this is not all bad. The protection of the customer's investment, and the product's potential for long-term usage are two attributes of a quality product. The name of the game in today's global, competitive environment is the offering and support of quality products and services. This certainly includes the design and use of network management protocols that define how we control our telecommunications systems.

More specifically, however, a protocol is a previously agreed-to procedure by which two or more separable entities pass control signals and information between themselves. These entities are, by definition, equivalent in terms of their position within the telecommunications system. They exist at the same level in the layered hierarchy of telecommunications, even though they communicate through other layers to reach each other. The way individuals start and end telephone conversations is an example of how protocols bring people together. The interactions between layers that make this communication possible are called the interfaces of the architecture. The telephone conversation is at one level, utilizing a protocol to allow the parties to communicate. This application (conversation) interfaces with at least one other layer that eventually passes the information back and forth through the physical layer by using interfaces to get between the physical layer and the ones above it. The physical layer carries the communications between the two ends through the switching and transport fabric of the system.

Multimedia network management requires the redefinition and development of new protocols necessary to support both the operation of the network environment as well as the management of that network. The principal criteria for these requiremnts are as follows:

(1)   Multimedia protocols should support variability in the delivery of media information based upon the variability of delay conditions as periodically measured during the ongoing operation of the system. In mode 1, increasing delays may warrant the manager to react by increasing the number of bytes per packet in order to smooth out such variable delays.

(2)   Multimedia protocols should be capable of conveying multiple information modes, such that operational traffic is defined by one mode, and network management information is defined by yet another mode. In mode 2, the media network elements may exchange status with the manager and commands back to the elements for those devices capable of receiving and reacting to management commanding.

10.2.4.1   OSI-Defined Protocols

The OSI model segments the communications process into a series of layers that have functional characteristics and for which standards of operation can be written [17]. These layers range in nature from the tangible or physical at level 1 to the creative or applications-oriented at level 7. The CMIP utilizes the OSI definitions for each layer to exercise network management tasks in each of these segments. There are several standards that have been initiated or are being developed that support the various OSI layers. The physical and functional structures of the OSI layers are described below. For network management purposes, a management system cannot be defined until some functional activity is present, at least at the physical or lowest level. In the meantime, standards are being produced that support functionality at each level and upon which network management development can be created. The standards work will support interface specifications between levels that will stimulate protocol interchange between equivalent layers.

Physical Layer (Layer 1) -

> Layer 1 provides the electrical or actual physical connectivity between network elements, including connectors, cabling, junctions, power, cable runs, etc., for the transport of analog or digital signals. Through these physical connections travel the analog or digital representation of data that are intended for relay from one point to another. The nature of the physical connection is sometimes selective as to which type of signal it will accommodate as well as the characteristics of that signal.

Data Link Layer (Layer 2) -

> Layer 2 provides error detection and control for the higher layers and is directed at digital signals packaged so as to maintain information integrity. This layer makes the physical layer reliable and provides a means of activating, deactivating, and maintaining the link status. It does this by adding flags to signal the beginning and end of messages, and by block check codes for error detection and, in some systems, error correction so that the layer 2 equipment at the receiving end can track and determine the status of the transmitted messages. In a voice system the equivalent would be signal conditioning and control.

Network Layer (Layer 3) -

Layer 3 is the means of establishing connections between stations. It provides the addressing and knows about the switching and transmission requirements necessary to pass data between entities. It is responsible for the routing and switching among the network nodes. It sends and receives control messages between itself and other nodes in the system to determine their status and to inform them of its status. Routing tables recognizing the appropriate area codes and numbers is rough analogy for this layer.

Transport Layer (Layer 4) -

Layer 4 has several functions, among which are the establishment of the end-to-end transmission paths where addresses become physical identities, assurance of the proper sequencing of data, prevention of data loss or duplication, and the optimization of network services. The establishment of transmission paths involves matching network addresses to actual device IDs (identifications) that are accessed by the users and the routing associated with those addresses [19]. Loss and duplication is avoided by use of sequence numbering of data packets, while optimization is enhanced by the use of end-to-end error detection and recovery capabilities. In voice and data systems, routing between two points may traverse many paths other than the most direct one for such reasons as load balancing or accommodation of equipment outages. Layer 4 provides this capability.

Session Layer (Layer 5) -

The session layer sets up the "conversation" between application layers. This layer also establishes the type of dialogue, e.g., 2-way simultaneous or 2-way alternating, and it provides some measure of recovery for the retransmission of data if problems have occurred between certain checkpoints. Specifically, if a communications problem occurs during an interaction, layer 5 will assist in the prevention of data loss and provide services needed to restart and eventually stop the task. It is roughly equivalent to saying "hello" and "good-bye" in a voice phone conversation.

Presentation Layer (Layer 6) -

The presentation layer is concerned with the syntax of data exchanged between elements and the format of that data, e.g., ASCII versus EBCDIC and so on. Additionally, it provides data and screen formatting service, and data compression as required. Many of the tasks of security control are exercised at this layer. The human conversational equivalent would be two parties speaking in the same language to each other.

Applications Layer (Layer 7) -

The applications layer is the realm in which application processes operate, and it establishes access to the environment for the purpose of exchanging data. The application processes may be spreadsheet programs, database management interfaces, message processing, etc. The utility of the system is expressed at this level. At this layer sign-on procedures, password checking, and system query and update control are also handled. The functionality that resides in this layer includes the creation of meaningful words and ideas, as well as analytical calculations. In other words, this layer relies upon external interfaces using human or human-defined expressions.

In the OSI model, communication between peers or equivalent layers of the model is accomplished by protocol, exercising communications paths between cooperating vertical layers, and is facilitated through one or more interfaces. For example, presentation layer communication is accomplished through the exchange of display control, formatting, and encoding information between nodes and is the responsibility of layer 6. The mutual understanding of format information between communicating nodes allows them to converse in a known manner. This communication exchange between the 6th layer at 2 different nodes is gated by interfaces at the layer boundaries.

The 5 major functions defined for network management in the ISO environment will be discussed later, but for reader orientation there is a rough correlation between the ISO layers and the these functions that will be provided here. These are: accounting - layers 4 through 7; security - layer 6; configuration management - layers 1 through 4; fault management - layers 1 through 7; performance management - layers 3 through 10.

The relay of information exchanges through the layers of each node and from one node to another is accomplished by appropriately packaging and stripping the contents of each transmission. This assures the integrity of information as it passes through each layer (much as each layer of skin in an onion protects each succeeding layer) and assures the proper regulation of that information. This process is called the *interface activity*. The way in which information is packaged for transmission represents the mutual understanding or protocol implemented between nodes.

There are, generally, two types of protocols, one being signaling and the other being message-oriented. The signaling-oriented protocol sets or raises and drops signal levels in order to notify interfacing equipment of the changes in the status of the unit making such changes. The message-oriented protocols set and communicate status or commands by passing coded status messages or commands back and forth. The OSI layers have varying numbers and types of protocols that have been or are being developed in order to serve the appropriate functionality at each of these levels.

10.2.4.2   Network Signaling Protocols

Protocols, such as those embodied in standards such as RS-232 for serial transmission, RS-488 for parallel transmission, and RS-449 for high-speed serial transmission and with sophisticated features to facilitate management, have been developed and made operational for some time to support the physical layer's functionality. These protocols are of the signaling variety. There is a facility within RS-232 for requests and acknowledgments between devices, notably data terminal equipment (DTE) to data communications equipment (DCE) that use the technique of raising and lowering signal levels. Additionally, RS-449 has a facility, among other things, for local and remote loop-back testing. In this way, the DTE attached to the circuit has a mechanism for tracing connectivity between itself and the nearest modem or transmission interface, as well as the modem or interface at the destination. This provides an effective and semiautomated way for a local terminal or management system to test the continuity of the circuit attached to the terminal in question.

10.2.4.3   Packet-Oriented Protocols

There are several packet protocols that have received endorsemant by international bodies, among which are the frame relay and ATM protocols. Both of these are based upon the same 53-byte packet, 48 bytes being devoted to the information being transmitted and 5 bytes devoted to the headers. These protocols were supposedly developed to optimize the conveyance of these types of data for continuous operational demands.

10.2.4.4   Message-Oriented Protocols

Message-oriented protocols are those features of the system implementation that provide the formatting, transmission control, sequencing, reset/restart, interrupt handling, call setup, and error handling necessary for data interchange. These protocols are amenable for use in message-switched systems, because message protocols and message switches operate from the same set of criteria, namely, the formats that encase the data. Message-oriented protocols are used mainly in the top 6 layers of the OSI model. If two nodes are communicating as peers, they do so either directly, because they have protocol compatibility, or indirectly, because their incompatibility is mediated by an intervening point of conversion or translation, i.e., a protocol translator. It is this conversion that serves as an additional step or element in the system that must be viewed as a potential point of failure and as an additional monitoring requirement for network management.

Each of the message protocols is implemented via software or firmware (hard-coded, algorithmic functions burned into read-only memory located in the hardware). A complicating factor is the fact that messages are being created and passed dynamically during network operations, and congestion starts to occur as the traffic intensity increases. This causes the probability of certain undesirable conditions to increase, i.e., livelocks and deadlocks. A *livelock* is an accident in the operation of the network, such as collision or stoppage at some node, whose existence is shielded from or not known to the users. A *deadlock* is a condition that causes some formal impairment to the operation of the network. Its existence becomes known to the users in a very specific way, such as when the system stops responding.

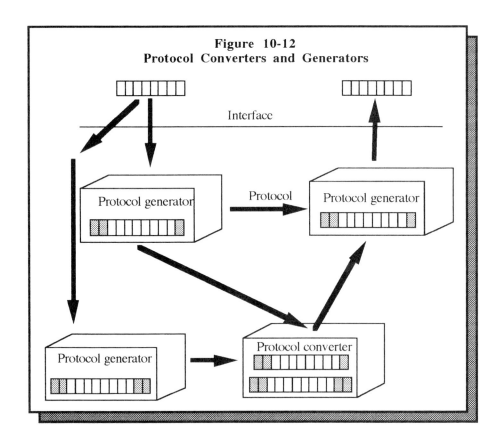

Figure 10-12
Protocol Converters and Generators

The notion of protocol compatibility and protocol conversion is illustrated in Figure 10-12. The existence of a dissimilarity in protocols or incompatible versions of the same protocol can create another failure point for the system that may have drastic impacts on system operation. These possible failures are related to the functions that the protocols perform. Figure 10-12 illustrates how protocols can communicate directly or indirectly via a protocol converter. Converters are functional elements that read an incoming data parcel from some unit, determine its type, reformat as necessary into the format of the destination unit, and retransmit the new data parcel to the required destination, with its original data content preserved intact. The user-specified portion of the data has thus been "containerized" into the new framework needed by the receiving protocol for routing and handling.

The next two sections introduce the reader to two of the better-known protocols to date, namely, X.25 and TCP (Transmission Control Protocol). Protocol utility and conversion issues are highlighted when reviewing such diverse protocols. On the one hand, X.25 is relatively simple but not as robust as TCP in handling duplications, time-outs, sequencing, and the like, all of which are needed for a protocol operating under difficult conditions. A brief discussion of each brings out the significant differences that highlight problems inherent in making conversions between the two, partly because of the differences in needed parameters and partly because of needed information requirements.

10.2.4.5   X.25 Data Communications Protocol

X.25 is itself a layered network communications protocol that encompasses the first 3 layers of the OSI model [19]. Two other protocols are used in the definition of the first 2 layers of the X.25 network protocol, X.21 for the physical layer, and Link Access Protocol (LAP) for the link layer. X.25 was developed originally to preclude different countries in Europe from developing their own packet systems, thus its creation made interoperability between countries possible. X.25 sets up a permission-driven session between originator and receiver. A virtual circuit number is assigned to the circuit over which the conversation is established, and a simplex (1-way) to full duplex (2-way simultaneous) interchange ensues. There are 2 basic types of packets in X.25, the hand-shaking packets and the data packets. X.25 also has an acknowledgment bit so that, if it is used, X.25 can exercise end-to-end control; therefore, X.25 acts at that point as if it is operating as a transport protocol as well.

This protocol has a header space of 3 bytes, so future expansion to accommodate greater sophistication is quite limited. It requires 2 addresses in order to initiate a virtual circuit, 1 for the remote destination terminal and 1 for the packet network itself. On the other hand, X.25 is simple and has low overhead and comparatively narrow bandwidth requirements as a result. For these reasons, it or some variation of it may have a future as the protocol of choice for such applications as packet voice and possibly video. Figure 10-13 illustrates the layout for the X.25 protocol format.

10.2.4.6   TCP Data Communications Protocol

The TCP protocol format is shown in Figure 10-14. TCP was designed to deal with noisy networks and is a testimony to the saying that necessity is the mother of invention [16]. The DoD commissioned the development of this protocol because the DoD realized the need for a system that could provide battlefield robustness; this was done in spite of certain international initiatives underway that were fostering the development of packet protocols. Using the assumption that the transport environment will be inherently noisy and disruptive as a criterion, the protocol was designed with several features to help meet the requirement for successful data transfer. The TCP transport station accepts user messages of variable lengths and assembles them into packets of no more than 65 kB for transmission.

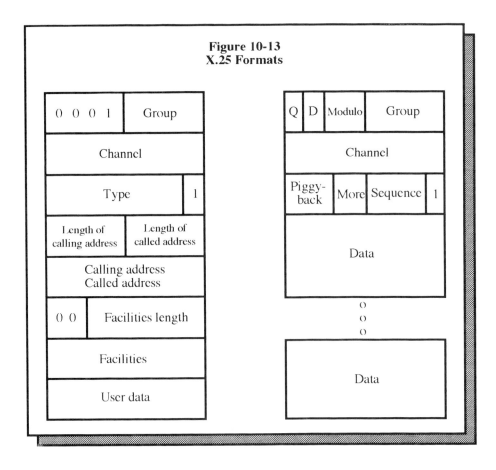

Figure 10-13
X.25 Formats

TCP also uses sequence numbers to assure that the same number of packets is received as were transmitted and are also used to facilitate reassembly of the packets into the proper order at the receiving end. Acknowledgment is handled by a piggyback field that returns the acknowledgment, using a return message rather than generating a special message for that purpose. A time-to-live field provides a duration to the life of a packet that may have gotten sidetracked. One of its important features is flow control that uses a window to tell the originator how many bytes may be sent beyond the current acknowledgment. Type of service requests, allowing the user to specify speed and reliability requirements to the network, are also accommodated in TCP.

Figure 10-14
TCP Format

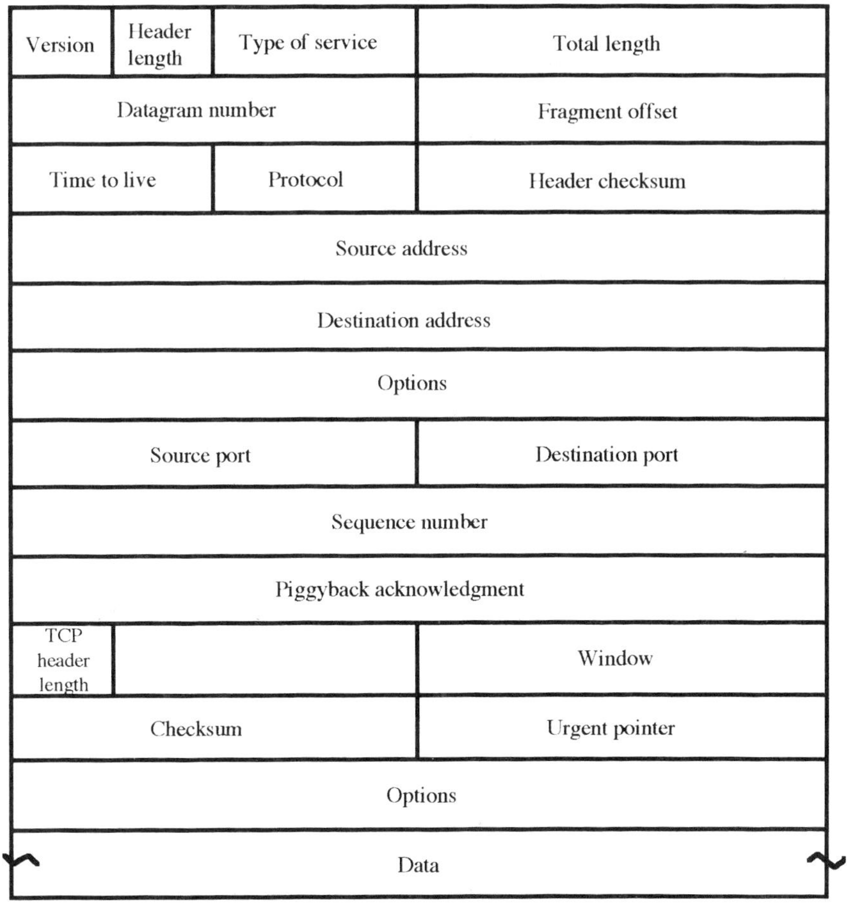

Version	Header length	Type of service	Total length	
Datagram number			Fragment offset	
Time to live		Protocol	Header checksum	
Source address				
Destination address				
Options				
Source port			Destination port	
Sequence number				
Piggyback acknowledgment				
TCP header length			Window	
Checksum			Urgent pointer	
Options				
Data				

TCP is suitable for the operation of LANs, WANs (wide area networks), and MANs (metropolitan area networks). Since bandwidth is not much of a problem for these technologies, TCP may be an appropriate tool. On the other hand, X.25 is utilized mainly in telephone environments, where bandwidth is more costly and the primary usage is to transfer large files. The flow control, sequencing and deduplication facilities of TCP may make it hard to replace as a robust messaging protocol.

10.3   Access to Network Management Information

It is axiomatic that network management can be performed only if information about the operational network is available to the network management system. There are two major issues confronting network management in this context. One is knowing the formats and presentation forms that are being used to convey operational and status information, and the other is the availability of or physical access to that information for analyses. In other words, network information has to be present for, and decipherable by, the network manager element of the system [9].

There has been much work done in recent years regarding standards, not only for communications *per se*, such as RS-449 and X.25, but also for data repository and organization, such as X.400 and X.500. As work on the standards progresses throughout the OSI layers, the possibility for more effective network management increases as the standards provide the syntax and semantical references for understanding the various transmissions within the system [16].

Access to information is, and has been, a prime focus of attention for computer systems applications for many years. Because network management is so closely aligned with computer operations, there is every reason to believe that information access and information decoding will continue to be one of the driving forces behind network management strategies for the future. On the other hand, the standardization of information acquisition through access standards may eventually lend some stability to the overall interpretation of information and, more important, to its effective usage in deciphering the semantical fine points of network operational status.

The existence of multivendor environments in digital networks presents a challenge to the system integrator who must provide a unified solution to its customers [19]. The very nature of the differing equipment types involved in network operations has made it difficult to develop taxonomies of information needs for the technical management of systems because of the variability of technical information available or necessary. The myriad of vendors and technologies has also created something analogous to a mystery problem, where network problem solvers schooled in the sciences and technologies must be multidisciplinary experts in order to unravel the secret of the right framework for the network management systems to support the jobs at hand [16].

Given the two network management issues of data availability and data deciphering described above, there are seemingly only a limited number of ways that network management architectures can be implemented.

One of the choices for the business organization is to manage the network itself. From a business viewpoint, another realistic option available to management is to contract out the network management function to an operations and maintenance (facilities management) company. Such a company presumably has the focus, experience, and expertise to deal with the questions of data access and the interpretation necessary. There are many such companies available that specialize in technical management. This business option can be exercised with any of the architectures indicated in this chapter, and its cost is directly proportional to the cost factors involved with each architecture. The approaches indicated here are based upon the data collection problem, that is, how data can be accessed from another vendor's equipment. Because network management architectures are dependent upon the connectivity issue as a consequence of the multiplicity of vendors available in the marketplace today, the problem has two major components. The term "primary" as used here means that direct connections are made between the network manager and the operational element(s) concerned. The term "secondary" as used here means that connections are made only between the network manager and the subnetwork manager station. The following subsections describe more fully the context and meaning of multimedia network management.

10.4 Network Management Schemes

10.4.1   ISO Network Management Functions

The question now is what functions are supported within the CMIP environment. In the network management environment, all the functions are in some way related to all other functions, either directly or indirectly. The tentative network management functions performed through the CMIP and defined by the ISO are as follows.

(1) Accounting Function -

CMIP will perform the statistical data collection tasks associated with the ISO network. The accounting function is a clearing house for all the other functions because user connections, usage, specific program accesses, database occupancy, and system problems are all routed to this function for accounting and record-keeping purposes.

(2) Configuration Function -

This function provides the database necessary to track and
categorize network resources, and to maintain their connectivity
status. It also provides the eventual capacity to order
reconfigurations based upon outages or necessities of increased
performance loading. System connects and disconnects are routed
through this function for association within the database. In
addition, the total inventory, including spares, maintenance status
on specific equipments, and software release status, are retained in
this function.

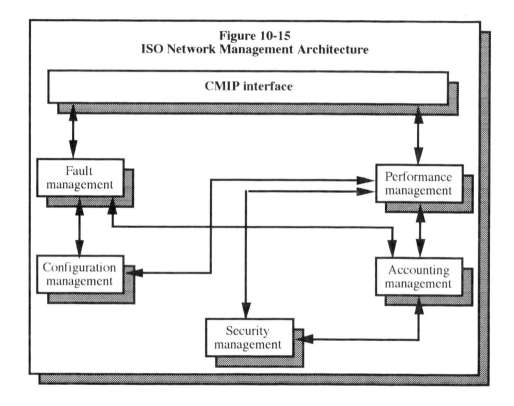

Figure 10-15
ISO Network Management Architecture

(3) Security Function -

System security is an essential element of the overall management
functionality. At this point, information concerning system user
identification, their privileges, and security access codes, e.g.,
passwords and identifications, is maintained. This function also has
the capability to provide audit tracking of user hookups that will
allow tracking of usage and, hopefully, flag attempts to penetrate
the system by unauthorized users.

(4) Fault Monitoring Function -

System errors are provided to this function for identification, analysis, and resolution, either manually or through some degree of automation. Fault indications are a subset of the total number of status messages provided to the system on a regular basis. However, because of the special nature of fault monitoring, a separate activity is provided to handle such information. Another important aspect of this activity is the collection of information useful in tracing network activity and the isolation of errors.

(5) Performance Function -

The system activities must be continuously monitored, either manually or to some degree automatically, in any system and in the least complex manner. The performance function provides this capability. Statistics are collected and analyzed so that trends toward malfunctioning that are more subtle than those black-and-white situations signaled in the more rigid fault monitoring function can be identified. Additionally, the performance function may be used to determine reconfiguration requirements and may be used for the planning of future system upgrades.

Figure 10-15 shows the general relationships for the network management functions and their interface with CMIP. All functions are highly interrelated and therefore must be integrated for appropriate and meaningful functioning.

10.4.2   Simplified Network Management Protocol (SNMP)

10.4.2.1   SNMP Description

The Simplified Network Management Protocol (SNMP) gained wide acceptance through the efforts and work of Stan Ames at Mitre, who, while managing a procurement activity, recognized the need for "built-in" network management capabilities in the systems he was then procuring on behalf of the U.S. government. The early project, Simple Gateway Management Protocol (SGMP), which stemmed from such early perceived needs, was the forerunner of SNMP. The refinement and development of SNMP as a specified protocol was the work of Jeff Case at the University of Tennessee and others. It was based upon TCP/IP, which was already a set of networking and transport protocols for the DoD that had been developed to deal with the noisy and unreliable data networks expected in military operational environments.

TCP/IP was used because of its ability to provide end-to-end transport
functionality, as opposed to handling just localized network transport. In other
words, TCP/IP has the capacity to establish and assure transport of data across
multiple network interfaces, not just within one local network. Later, the ISO
model specified this transport capability as the 4th layer of the ISO model,
thereby matching the TCP design goals, and the network capability of IP
(Internet Protocol) as the 3rd layer, thus paralleling its objectives as well.

The TCP/IP ensemble actually encompasses only 2 of the ISO layer
functions. *Transmission control protocol* is a transport-layer protocol that was
originally developed by the Defense Advanced Research Projects Agency
(DARPA) to handle "unreliable" packet transmission. This means that this
protocol would assure that packets were delivered to their destination and that
no loss of packets would occur. Along with TCP, a new 3rd-layer protocol was
developed called Internet Protocol, that was charged with routing packets
around the network and handling functions such as fragmentation of
datagrams and their reassembly at the destination end. These 2 protocols
together were very appropriate as the backbone of a network management
standard that was to emerge, because together they provided reliability in
difficult conditions and were aimed at assuring delivery at the receiving end.

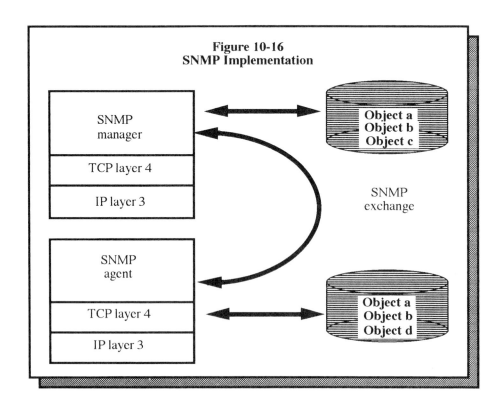

Figure 10-16
SNMP Implementation

10.4.2.2   SNMP Architecture

Figure 10-16 shows the SNMP concept of network management. The protocol is exercised via a set of agents and a manager [17]. The protocol is a very straightforward and elegant scheme that relies on low-risk implementation and understandable functionality, as well as interactions between the two main elements of the system. The manager holds what are called objects or functional modules that perform specified tasks at regular intervals, or perform other tasks on demand as ordered by the manager control element. The manager, hopefully, contains all the objects that are to be executed within the network over any given time frame.

Each agent, in turn, contains at least those objects that it needs or is capable of performing. Sometimes, an element may contain an object that is not resident in the manager, which means only that the object is not executed. Alternatively, the manager may contain objects that some element may not contain, which means that this object is not executed at that agent unless some downloading capability is provided to that system. The key, however, is that the system is set up to provide a building-block environment that can be added to or subtracted from as necessary for the management of the activity.

The objects that reside in the network manager and agent stations are part of the overall management information base (MIB) and respond to the various commands available to the requester, such as "get," "sett," "trapd," "nextx," and so on. As each command is sent and received, the appropriate information is retrieved and the response module (object) relays the answer back to the requester. Objects might, for instance, provide status reports on a packet switch that would result in such reported feedback as numbers of packets sent and received over some period of time [17].

10.4.3   CMIP versus SNMP

The comparison between the two systems is like looking at two sides of the same coin [16]. Unlike CMIP, SNMP services only 2 layers since it rides atop TCP/IP. SNMP also has "verbs" with which the user solicits information or directs action, but so does CMIP. Network management based on the OSI model relies on the entire 7 layers of the OSI model for organization, while TCP and IP rely upon the 2 layers of the OSI model known as the transport and network layers for their structure and architectural support. Additionally, CMIP contains a richer database than that contained within SNMP. In SNMP this database facility is referred to as the *management information base* (MIB).

CMIP, also, will eventually have more "verbs" and therefore be capable of greater remote control over network elements than SNMP. Theoretically, CMIP will have better confirmation and error-messaging features than SNMP as well, so that users will know that commands are executed or, if they are not, why. The key, however, is that SNMP provides network management now and has actually acted as a prototype for CMIP. Both ISO, representing CMIP, and the Internet Engineering Task Force (IETF), representing SNMP, are developing security standards, the OSI standard being X.800 [20]. Security seems to be a

weak area for both at present, but this flaw is being worked on and will no
doubt be corrected, in time, in both protocols [16].

The major difference between CMIP and SNMP is that the network
management of each is present only to the extent that layers are defined for
each. If one were to define a system with 3 layers, then it would make sense to
discuss network management only in these 3 layers. It is difficult to discuss
what has not been conceived nor described. Thus, the situation with SNMP is
that it is somewhat trivial to discuss 7 layers relative to this protocol because it
is not based upon 7 layers. It treats 2 layers at most but pretends to do nothing
else. In so doing, SNMP provides more breadth in these layers than does
CMIP. On the other hand, if network management is defined and based on 7
layers and not 2, then SNMP falls short insofar as that definition is concerned.
The question to be answered eventually in the marketplace is whether CMIP's
depth is necessary or warranted. In many situations, no doubt, it will be highly
desirable, especially for safety or financially critical environments.

10.4.4   Manufacturing Automation Protocol (MAP)

Another protocol that has received much attention in recent years is the
Manufacturing Automation Protocol (MAP) [7]. The impetus for this protocol
development was the high cost of interfacing certain kinds of specialized
machinery used in various types of industry, notably the automobile industry.
This protocol was initiated by General Motors at a time when the cost of
interconnecting different types of incompatible devices was a software
development expense of very high order. Work got under way in 1980 to
develop a standard for manufacturing machinery, which in the case of General
Motors amounted to over 40,000 different types of devices that required
software interfaces. This protocol, while based upon the ISO model, was
tailored for specific problems, such as a process control environment, as cited
above. The first MAP specification was published in 1982 and revised in 1990.

In the MAP system, the user is actually an application process, which can
assume the form of any task from a simple program to complex operation that
functions as part of a piece of machinery. The user, or process, is at the
applications layer and has two different kinds of services available, these being
common applications service element (CASE) and specific application service
element (SASE). These services may also refer to different kinds of design or
cutting tools. MAP provides a directory of all such CASEs and SASEs in the
network system as well as a network management application. MAP also uses
an OSI-defined specification called file transfer and management (FTAM). This
standard provides the services and support protocols that allow applications to
create and delete files. FTAM is used in MAP as a means of providing
association and connection services.

The network management function defines two components, one called
a network manager and the other called a network manager agent. At each
node there exists a network agent that all the layer (functional) managers must
go through in order to communicate with each other, e.g., modem manager to a
packet manager. These network agents provide an interface to the single

network manager application that monitors all the agent traffic and, therefore, knows about all the activity between the layers.

The network manager sends requests to the agent for values of parameters or instructions to modify a parameter. The agent may, however, initiate a message to notify the manager of outages. There are 11 abnormal events defined in MAP, such as a protocol violation, unsupported or impossible options, and operational aborts by specific machines that cause an event to be generated and notification provided to the network manager. If the manager does not have an established association with the agent at the time of the event, the event will be ignored. The architecture for this protocol is represented in Figure 10-17.

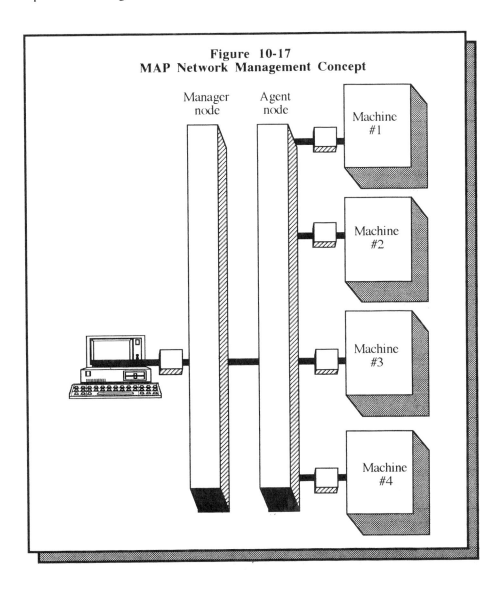

Figure 10-17
MAP Network Management Concept

10.4.5   Systems Network Architecture

With 80 percent of the *Fortune* 1000 companies running Systems Network Architecture (SNA) on their mainframe computers, it appears that IBM will maintain a significant share of the network management market with its Netview product. Even though this is a proprietary network management package, both SNA and Netview are *de facto* standards that cannot be ignored. However, eventually both will have to become compatible with CMIP and SNMP. IBM and third-party providers are developing other packages to provide compatibility between SNA and the other network management facilities. Open architecture is taking on a special meaning not only for IBM but for all other vendors as well. The trick has been to get everyone to adhere to one version of what an open network architecture is.

Technical Note 10-1
Systems Network Architecture

IBM has long pushed a proprietary architectural concept known as Systems Network Architecture. It is based upon many protocols that support the IBM hierarchial approach to the control and management of the various network resources. A closed, or proprietary architectural concept tends to limit vendor support. On the other hand, an open architecture approach has been the best way to foster compatibility between hardware and software facilities. This approach is taking on a special meaning not only for IBM but also for all other vendors as SNMP and CMIP garner increased support from users and developers alike. The challenge for the proprietary products has been to convince others to interface with their systems and products, thus opening up their architectures to additional standards and a wider range of equipment, such as SNMP and CMIP.

Netview and other proprietary systems such as AT&T's unified network management architecture have, to this point, targeted a subset of the total array of telecommunications products on the market, but this position is changing with the more inclusive standards discussed above. For instance, IBM's Netview operates within the SNA umbrella and is a program running on an IBM central computer that monitors IBM or IBM-compatible, system-wide functions. The basic change for these proprietary systems will be to accept CMIP and SNMP messages and to convert them into a syntax compatible with these proprietary products. Finally, these systems must be converted to interpret these new types of messages. Alternatively, some third-party vendors are also developing packages that accept Netview messages, for example, and convert them into a different syntax for use in other network management systems.

Netview operates within the SNA umbrella and is a program running on a central computer that monitors system-wide functions. The basic approach for IBM is to accept CMIP and SNMP messages and to convert them into the Netview syntax. Some third-party vendors are also developing packages that accept Netview messages and convert them into a different syntax for use in other network management systems. Technical Note 10-1 provides additional background for IBM's SNA.

10.5 Operations of Protocols

As explained above, a protocol is analogous to the procedure used in making a phone call or the steps in executing an administrative process [17]. Take, for example, the process involved in making a phone call. The initiator accesses the network by picking up the handset. Then he or she provides the network the information necessary to make the connection to the receiver. The called number is used by the network to develop the physical contact to the destination. The system signals the receiving end that a call is coming through. The receiver accesses the system by picking up the handset. Finally, the receiver acknowledges that he or she is on the line and the caller requests identification, on occasion, to assure that the correct party has been reached. Sometimes the caller knows the receiver's voice.

The implementation of a data communications protocol is much the same as the scenario described above. The calling process accesses the system via some request that initiates the opening of an object that in turn carries out the communication request. The user specifies, among other things, the destination and content of the message. The system is responsible for the routing of the message or message parts, using lower layers of the OSI model. The message is provided to the receiving end, where the recipient is alerted of its arrival and interprets the message sent. In most cases the parties would set up the session, or conversation, prior to the transfer of data.

The scenario described above is carried out automatically millions of times each day by thousands of networks. Unfortunately, there are many ways in which even the best and most robust (self-healing) protocol can fail. Hence, there is a need for network management even in a perfectly connected and very reliable network. It is interesting to note that 28 percent of all errors among the total of network failures occur in the top 4 layers of the OSI model, and that a great part of such problems are attributable to protocol problems. Thus, protocols are not infallible. They present potential sources of failure in a network, especially a heavily loaded network where contention becomes an issue. When different protocols have to work together, the potential for failure is magnified.

10.6 Economics of Protocols

It is reasonable to suspect that as one moves from lower layers in the ISO model to higher layers, the variety of options and features used in protocols increases.

For each layer identified and involved in an existing or proposed protocol, there is a potential for increasing the complexity of operation and therefore the chance for error. Each feature of a protocol and each optional implementation of a protocol increases the chance for error. Engineers have long known that the potential for error or operational breakdown is related directly to system complexity, whether that complexity resides in its parts or its functions.

It sometimes takes many years to build confidence in systems and/or methodologies that support a high degree of complexity or varieties of features. This may be why users today are interested primarily in simply assuring the integrity of connections [19]. Only in situations of life-and-death struggles, such as illustrated in the 1991 war in the Near East, are new technologies tried out in advance of desired schedules, or exhaustive test scenarios to back up the anticipated results. Fortunately, such situations do not occur often in network management environments.

The system user can expect to invoke many new chances for error as new functionalities are brought online. The KISS rule, "Keep it simple stupid," is sometimes the best strategy. However, the advancement of capabilities can be furthered by the use of redundant components or parts that support increased functions while at the same time providing the basics for backup purposes. Additionally, the use of gates or gateway functions that are tried and subsequently used as the conduits for features and functions that may be added later is another way of minimizing chances for error.

Using the tried-and-true approach, one function, such as the user interface, may be developed and tested rigorously to assure reliable performance, i.e., consistent results, and valid performance, i.e., relevant truth or what is appropriate for the use at hand. As each new feature is delivered, it is subjected to the requirements of the interface and, where deficient, is reworked to comply appropriately. In this way the system is protected from multiple failures and focuses on one point of possible failure. For software implementations, a single interface does not affect availability in an engineering sense because there are no moving parts that may be subject to failure over time. For electronic or mechanical implementations, redundancy is the protection against lower availability of the functions.

Quantifying the economic consequences of failed or partially failed protocols is a difficult activity. In a very simplistic way, each and every task performed in the protocol can be assigned a probability of success. If each of these are multiplied together, eventually we will arrive at a point where the odds of a failure are very good. Additionally, the chances of failures are enhanced as the congestion in the environment increases. Such an increase in error probability appears to follow the same curve as the increase in arrival rates of objects into the system.

Summary

The network architectures that dictate and control the implementation and use of multimedia systems are key to the physical integrity and cost-effectiveness of the target multimedia network. The management of data networks has been the focus of intense standards work for over 10 years, and it appears that multimedia networks will be able to take advantage of the work in this area in so far as the physical control of the systems is concerned. As discussed previously, multimedia applications and the products that they generate is judged mainly more qualitatively than quantitatively. As a result, the applications level will have to provide measures of the quality and therefore the management of the multimedia network. This chapter has outlined the similarities and therefore the parallel utility of current network management efforts that will serve the multimedia network well in the future.

References

(1)   Bonafield, C., "Router-to-Router Spec Advances OSI," *Communications Week*, March 19, 1990.

(2)   Breidenbach, Susan, "Customization of SNMP Products Obscures Benefits," *Network World*, April 23, 1990.

(3)   Carter, R., "MAP Communications and Network Management," *Proceedings of The Third Annual Artificial Intelligence & Advanced Computer Technology Conference*, Long Beach, CA, 1987.

(4)   Desmond, P., "Standards Push for TCP/IP Miffs OSI Backers," *Network World*, June 18, 1990.

(5)   Ezerski, Michael, "OSI Network Will Evolve in Phases," *Computer Technology Review*, May 1989.

(6)   Goyal, S. K., and R. W. Worrest, "Expert System Applications to Network Management," in *Expert System Applications to Telecommunications*, Liebowitz, J. (ed.), John Wiley, New York, 1988, p. 1.

(7)   Harris, C. J., and I. White (eds.), *Advances in Command, Control and Communication Systems*, Peter Peregrinus, London, 1987.

(8)   Henning, W., "Bus Systems," *Sensors and Actuators A*, 210-27 (1991), pp. 109-113.

(9)   Hoffman, M., "Technology Profile: Neural Networks," *Techmonitoring*, SRI International, July 1991.

(10)   Jackson, Kelly, "TCP/IP-OSI Clash Goes ON," *Communications Week*, October 1989.

(11)   Lemmon, A., *Marvel - A Knowledge-Based Planning System*, GTE Laboratories, Internal Report, 1986.

(12)   Marney-Petix, V. C., *Networking and Data Communications*, Reston Publishing, 1986, pp. 50-90.

(13)   McGrew, P.C., *On-Line Text Management: Hypertext and other Techniques*, Intertext Publications, McGraw-Hill, New York, 1989, pp. 40-50.

(14)   Peters, Lawrence J., *Software Design: Methods & Techniques*, Yourdon Press, Englewood Cliffs, NJ, 1981.

(15)   Puttre, Michael, "CMIP: The Next Phase in Network Protocols, Eventually," *MIS Week*, June 4, 1990.

(16)   Puttre, Micheal, "SNMP Eliminates the Need for Network Management Gurus," *MIS Week*, May 28, 1990.

(17)   Rose, Marshall T., "Taking the Helm," *Mini Micro Systems*, November 1988.

(18)   Shandle, J., "Price Tag for Faulty Nets: $3.5 Million," *Electronics*, October 1989.

(19)   Gibson, S., "IBM's TCP/IP Effort," *Communications Week*, June 18, 1990.

(20)   Turner, S., "The Network Manager's Compendium of Standards," *Network World*, April 15, 1991.

11

Multimedia Network Architectures

Chapter Highlights:

The architectures of multimedia systems emphasize packet communications, quality measures that are needed to manage their operations, and shared bandwidth to facilitate 2-way communication sessions between humans, or in some cases between humans and computer systems. The understanding of these network configurations, access to network information, the protocols, and associated standards all combine to frame the constraints and limitations that indicate the architecture of any particular system configuration. In this chapter, the reader will begin to understand these capabilities and limitations aligned with the network management problem and, in particular, the multimedia network management problem.

11.1   The Special Nature of Multimedia Networks

Unlike data networks, multimedia network architectures are driven not only by their connection requirements, but also by their bandwidth, information delivery

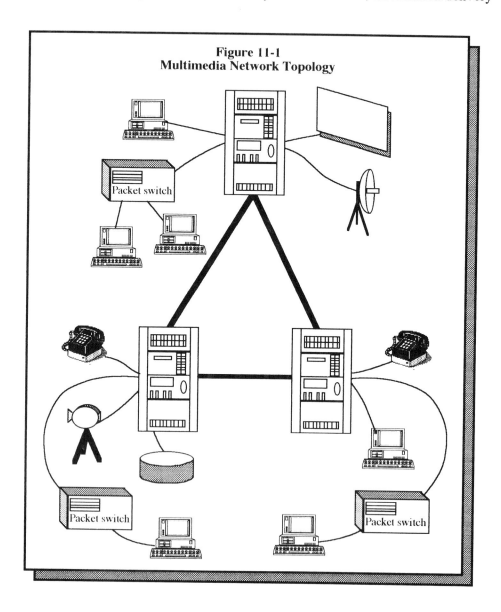

Figure 11-1
Multimedia Network Topology

delay, media access, and quality requirements [5]. These issues have been identified elsewhere in this book, but they will be tied together here in this chapter. First, we will review the basic features of the networking options available to us and explore where we must develop new technologies or new techniques to take advantage of the promise of multimedia, which is:

To become an extension of human capabilities in order to reach as far as necessary to communicate with other humans in a humanlike manner.

Network managers can expect broadband packet-switched networks and LANs to be more and more available as the twentieth century draws to a close. The continued use of twisted copper pairs and coaxial cable will allow short runs from the sources to the concentrators and from there to broadband facilities where addressing and delivery is determined by packet headers. Figure 11-1 is a conceptual view of how this will be accomplished. The typical network will consist of broadband facilities that will attach to a variety of devices, which will, in turn, provide user access to the network and eventually other users.

11.1.1   Classification of Generic Networks

There are three different basic types of networks in existence today [3]. These are:

(1) **The Message-Switched Network -**
Message-switched networks are those that take complete sets of information and route them according to some defined scheme. The messages are complete in and of themselves and not fractured so as to create a need to track the number and order of different parts of the transmission.

(2) **The Packet-Switched Network -**
The packet-switched network is one in which the information is broken into standard-length pieces, known as packets, so as to create less congestion on the network when numerous messages are being transmitted from one point to many or between many points. The packet makes the management of the information interchanges more predictable, thus assuring more reliable estimates of delay, and therefore, the system is better able to provide video and audio services in a digital environment.

(3)   The Circuit-Switched Network -
The circuit-switched network is one where the connection between
the 2 parties concerned, whether they be both human or human and
machine, is established by a physical linkage between those
concerned. This is usually accomplished by an electronic switching
system where every subscriber in an area is connected to the
switch. Subscribers who are not in the local area are reached by
high-capacity trunks that carry the connections between switches.

The message network carries entire messages between points to/from the
communicators. An example of a message-switched scenario is the transmission
of a movie from a video headend to the subscriber(s). This type of network,
however, is used commonly for short messages (datagrams), on a framework of
fairly constant intensity, or loading conditions. The packet network splits up
messages, such as files or other lengthy communications, into pieces or
segments of manageable sizes. This segmentation minimizes waiting time within
the transport system and supports the passage of these packets between the
sender and the receiver with greater efficiency, because of the packet system
facilities for handling these messages.

With the packet approach, the implication is that all pieces will arrive at
the destination within a short period of time of each other, and that there is
some way of reconstructing the entire message upon arrival. This type of
network is used for lengthy messages that would otherwise clog up the system
if they are sent complete and thus delay other messages. The packet system
may allow for messages of different origins to be interleaved and/or sent to the
destination via different routes if necessary.

The message-switched and packet-switched networks make their
connections via logical switching. This means that routing and transport are
accomplished by addresses, whereas physical switching is accomplished by
switching devices that make physical connections between the senders and
receivers. The packet-switched network is connectionless-oriented, because all
nodes are physically connected at all times, although not necessarily
communicating with each other. Communication occurs only when specific
pieces of information addressed to some specific node arrive and are recognized
by the receiving node.

The circuit-switched network allows individuals to capture a part of the
system for their own private use for the duration of the voice conversation or
computer session. This approach is used in the public telephone network to
provide multiple, simultaneous conversations. The circuit-switched network is
connection-oriented, because contact is made by the physical making and
breaking of network connections. Table 11-1 shows some of the various
standards used in the OSI layers.

Table 11-1 Important Standards for the OSI Layers	
X.400 (message handling system) ISO 9040/41 VTP (virtual terminal protocol) TP (transaction protocol) X.500 (directory protocol) Network management protocol ISO 8571 FTAM (file transfer access mgmt.)	Application layer
ASN. 1 ASCII	Presentation layer
ISO 8326/8327 (session service)	Session layer
TCP (transport control protocol) CCITT X.75 NCP	Transport layer
CCITT X.25 IP (internet protocol)	Network layer
IEEE 802.3 Ethernet (STP/UTP) IEEE 802.4 MAP token ring IEEE 802.5 LAN token ring IEEE 802.6 MAN IEEE 802.9 LAN/ISDN integration ISO 3309 HDLC	Link layer
ISO 9314 FDDI EIA RS-232 X.21 (X.25) EIA RS-449	Physical layer

11.2 Architectural Configurations

11.2.1   Architectural Problem Areas

Experience has shown that over a third of the network management problems in any one particular network are caused by the physical connections within that system. Also, the more the connections, the more likely a system is to experience network problems. These are very powerful statements, and they alone certainly justify a considerable amount of effort just in assuring the proper physical control over network elements as well as their proper monitoring. Such potential physical connection problems include the wiring, connectors, terminations, etc., that make up the physical connections between network elements. These physical items also contribute significantly to total costs, depending upon how they are configured to interface with each other through the system architecture.

Each physical termination provides a new possibility for failure and is a possible point to which the network management system must look in order to resolve outages. Each physical part of the system can be made to be essentially "bombproof," but such a level of scrutiny comes at a price. The point here is that even at this lowest level of system functionality, cost trade-offs become critical. At this first level of investigation, operational and maintenance costs will have to be traded off against capitalization costs, each being a major contributor to the annual costs of the total system. As the maintenance requirements increase, costs begin to rise rapidly, and accommodation for possible new technologies or new equipment, at the very least, must be addressed.

The focus here is upon the connectivity among the various network elements that, in turn, influence the total system functionality. Nowadays, however, very few total functional system failures occur. An example of such a total failure might be a software malfunction that terminates or erroneously reroutes communications traffic, such as the telephone system outage that affected the Eastern Seaboard of the United States in 1990 as a result of an event known as the Hindsdale fire. Such malfunctions, while disastrous, are fortunately few and far between. For the most part, systems fail in isolated areas or in limited ways, and their failures are many times physical in nature. This why it is even more important to understand the totality of the pieces that make up the system configuration. Additionally, it is important for the system to be monitored from as many points as possible.

For purposes of this discussion, we can categorize failures as being derived from individual physical components, the total physical system, individual functional components, or the total functional system. The total physical system is unlikely to fail all at once, for any number of reasons, not the least of which is that redundancy or reliability will tend to preclude such an event. The total functional system is also unlikely to fail all at once for the same reason. This leaves the subsystem components as being likely to cause problems, with outward functional symptoms masking the physical problems that might be the real cause.

Functional failures at the component or subsystem level most likely will manifest themselves as more than one malfunction. As the alarms for some malfunction spread along the signal paths, the isolation of the problem becomes more and more difficult. Thus, if the monitoring points of the various segments along the signal path can be made more frequent, the opportunity for quick detection and isolation can be enhanced. This chapter addresses these last two categories, in particular, from a technical as well as a financial viewpoint.

11.2.2   Architectural Diversities and Requirements

Functional requirements, or business drivers, and business cost considerations drive the architectural definitions of systems. This is especially true of large-scale systems, where small variations in cost can drastically affect margins and slight differences in functional features can, in turn, significantly affect costs. Take, for example, a community area TV (CATV) system versus the typical voice telephone network, or a direct broadcast system, e.g., as exemplified by a satellite system. All 3 are focused on communications in varying ways. Each is also driven by its respective business objectives, which may also drive architectural designs. Yet all are distinct in their architectures: the first being a tree configuration, the second being a switched system, and the third being a true star system. Of course, additional factors, such as the markets addressed and the evolution of these systems have also contributed to their respective diversities.

Another form of diversity can be found in more simplified communications systems. A point-to-point system used by two people to communicate is probably the simplest of systems. In such a system, voice contact can be realized by nothing more than a telephone connection. These same two people can also communicate in a slightly different way even though they are still connected directly to each other, this being via data representation of words and images. Such a system can be effected as an adjunct, or extra feature to the voice system, using integrated voice and data circuits. For distance communication the modem is employed. The modem transforms pulsed information representing words and images into a form that can be transmitted over great distances. Now the two communicators have two methods of exchanging ideas and thoughts, as typified by data communications equipment. Other data communications methods include video and an integration of all of these modalities. The range of possibilities for thought transfer is illustrated in Figure 11-2.

The role of digital technology has increased significantly in recent years. One of the primary reasons for such enthusiasm is the ability of digital technology to support and represent all forms of human communication. If the two ends of a circuit are connected with the proper equipment, the pulsed information can be used to replicate images and/or graphic representations of meaningful information as well as voice transmission, using facsimile or other methods. Thus, the pulsed information or digital data is now useful for picture as well as word transmission. If the system is fast enough to transmit and update these images at about 30 times a second, the viewer may experience the feeling of seeing motion or, as it is referred to, video. Finally, if the communicators can send and receive information in printed form, then they have realized the extent of human communication, i.e., voice, print, and video. All three modalities can be accomplished via digital data or pulsed transmission.

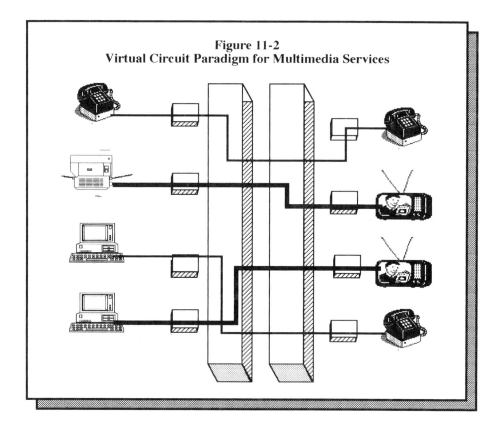

Figure 11-2
Virtual Circuit Paradigm for Multimedia Services

This extreme simplicity can now be complicated somewhat by a 3-way, or more, conversation or communication session among groups. At this point the question arises of how the information is conveyed and sequenced among the 3 participants. This can be done either by periodic access directly to a hard-wired (physically fixed) network or through some form of switch that sets up the connection to be transmitted over the air waves (wireless) or over a hard-wired network.

If a switch is involved in the communications session, considerable expense may be incurred in providing that capability, because the flexibility that may be required in providing circuit-switched service has to be paid for in the cost of the switch [3]. If a direct connection to some form of hard-wired network is involved in the setup, the cost of the interface and access to the desired destination may be limited to the hardware and software protocols necessary to make the logical connection and handle the conflict resolution. The need for multiway communication has led to the need for networking so that the expenses of point-to-point connections can be mitigated and the inconvenience of manual connections can be eliminated.

One of the 3 types of networks identified above is required for multiway communications. Each has its own advantages depending upon the application involved. All, however, have some features in common: namely, all three types of networks require some means of providing addressing capability and some way for establishing the connection between sender and receiver. The various configurations for making this happen are discussed below.

11.2.3   Specific Design Options

The primary reasons for variations in architectural configurations involve the cost and usage needs of the system. The trend toward distributed processing and database systems has given increased importance to the various network configuration designs, and as such their discussion is necessary here. The cost and usage requirements for a system can be analyzed in many ways, one of which is discussed below as the classic allocation problem. First, however, it is also important to understand the implications of the various architectural forms available. Each architecture has both desirable and undesirable points [1], as indicated in the following discussion and illustrated in the succeeding diagrams.

(1)   Star Network -

This configuration is based upon the distribution of information from centralized points to and from outlying destinations. This centralized distribution point may, in turn, be serviced from another centralized point of higher order. The idea here is that such a configuration has to deal with the switching of information to/from source points to any other receiver point or points. It is controlled by an operating system, or something equivalent, that arbitrates the actions of the hardware. The management of such a system in terms of control and problem definition is straightforward, because there are only so many points at which problems can occur or at which control can be exercised, namely, the transmit and receive nodes, the transmission paths, and the switching node. The telephone system in the U. S. is a connected star network that was developed using this architecture because of its relative simplicity. The early computer systems were all star configurations, with the central processor as the center of the star and the connected points being the terminals or attached workstations reliant upon the central unit for support.

The latest packet communications technologies, such as frame relay and ATM are based upon a star configuration at the local level with access to similar packet switches in adjacent districts. Star networks, relative to other network configurations, are generally thought to be expensive to build, inexpensive to operate, and moderately priced to its customers. See Figure 11-3.

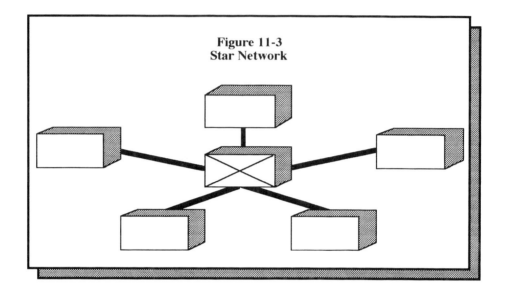

Figure 11-3
Star Network

(2)   Bus Network -

The development of computer systems architectures gave rise to the concept of a bus, or highway, approach to information transfer. Today, certain local area networks (LANs), such as Ethernet, exercise the bus configuration in a similar fashion. Under this architectural concept, the various elements of the system are in contact with each other, directly or indirectly, through a master control element, over a common electrical path. The bus configuration, shown in Figure 11-4, is used almost exclusively as the internal communication path for computer stations and terminals. This architecture has certain attributes, such as total information sharing, and certain restrictions which are managed by its operating system, such as the priority control of communication between elements. Priority control means that permission to use the bus is selectively allowed by the operating system based upon certain conditions, such as proximity to the central processing unit, as with the VME bus. The principal functions exercised by the operating system or master controller, are conflict resolution among the various competing elements of the bus. Permission to use the system is granted on the basis of a specific methodology, such as first come first served. This type of system is moderately expensive to construct, inexpensive to operate, and inexpensive in cost to the customer.

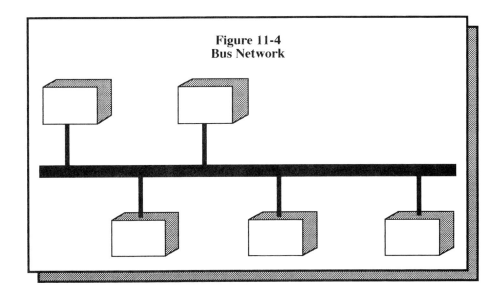

Figure 11-4
Bus Network

(3)   Ring Network -

Various approaches have been tried to solve the problem of how to
pass information among related elements most efficiently or reliably.
The ring architecture, illustrated in Figure 11-5, is one such
approach that has received much attention and wide acceptance,
especially for small or self-contained local area networks. In the
ring system, all elements communicate over the same path that
connects all other elements. There are presently two primary
approaches used to carry messages in such systems. In the token
ring approach, there may be a special control message for
communication control, called a token, that continually circulates
around the ring, stopping as required by the protocol at appropriate
nodes. When the token is captured by a particular node, that entity
has control of the network temporarily and is thus able to initiate
the transfer of data or a request to another node. When the second
node receives the token, with the data or request embedded, this
node acquires temporary control until it responds to the request or
releases the token unused. This type of approach works better than
the contention approach under high loading conditions.

The other ring concept is called the contention approach
and operates on the theory that all elements have permission to
access the system at any time. This may seem chaotic, because all
the nodes may want to communicate at the same time, thus
causing a lot of retransmissions due to the collision of messages
during operations and resultant retries. The simulation of such an

arrangement, however, shows that under controlled circumstances, this approach works very well. As a matter of fact, the contention method works well with low to moderate loading on the system, whereas the token ring method works best in moderate to high loading situations. The ring architecture is moderately expensive to implement, moderately expensive to operate, and moderate in cost to the consumer of its services.

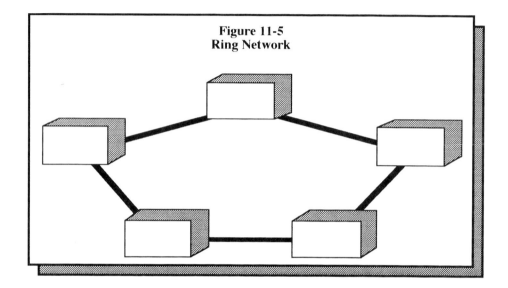

Figure 11-5
Ring Network

(4)   **Branched Network -**

A branched network is an architecture of convenience. It provides an efficient method of getting information from some principal point to all destinations in a broadcast-like fashion. The concept for branched networks is illustrated in Figure 11-6. Its configuration is very reminiscent of some type of hierarchical organization and can be thought of as exactly that. It mainly commands and controls through message passing or a rendezvous mode, where control signals are managed by relay from 1 layer to another or by common distribution points. This type of network is used extensively in the CATV industry to convey TV signals, because of this characteristic. It has not, however, flourished as a means of accomplishing 2-way communication, because the very nature of the technology encourages additive noise to corrupt any "upstream" signal or information content. Additionally, the cost of noise-suppressant hardware has made its use as a 2-way medium relatively uneconomical to this point, although several attempts to use it in this manner are ongoing. The branched network is, or has been, very inexpensive to construct, inexpensive to operate, and moderately priced to the consumer, thus far. It has been the network configuration of choice in the CATV industry, because of its low capital costs and operating expenses.

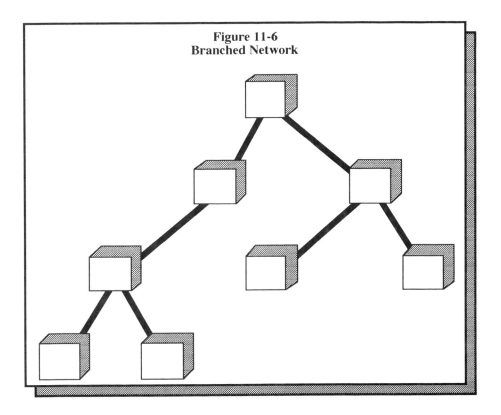

Figure 11-6
Branched Network

As networks grow independently due to geographic separation or because of increasing complexities, needs arise to integrate these diverse networks or to bridge communications between them for operational conveniences. Such devices are added between the configurations discussed above to support internetwork configurations. Such devices will be discussed in more detail below.

11.3 Functions of Telecommunications Protocols

Protocols perform several functions in a communications environment [4]. These are:

(1)   Parsing, or segmentation with subsequent reassembly

(2)   Encapsulation or formatting

(3)   Connection control or communication setup

(4)   Ordered delivery or message administration

(5)   Flow control or end-to-end circuit control

(6)   Error control or error handling

(7)   Multiplexing or interleaving

Not all protocols perform all of these functions, and some do not perform them very well, but an ideal protocol will have these features adequately implemented. Each of these functions is explained briefly below.

Parsing -

The protocol takes the message and breaks it into the pieces necessary for transmission. This process is performed first so that the total number of pieces, or packets, can be determined along with, in some cases, a sequence number assigned by the software to each packet and carried as a count in each data parcel that goes out.

Formatting -

Each protocol has its own requirements for placement of routing indicators, error checks, etc. These requirements differ among protocols in type and location. Each protocol takes the data piece previously parsed and places it in the data parcel frame along with the administrative information necessary for routing and handling along the way.

Communication Setup -

Conceptually this is a straightforward process in concept, which calls for the establishment of the communication interchange. The process consists of establishing the connection, transferring the data, and terminating the connection.

Message Administration -

The administrative information loaded into the data parcel is used to assure delivery of the information to its destination in the right order. Recall that the parsing process determines the total number and sequence number of each data parcel. This information is used to provide the means for reconstructing the sequence at the receiving end and determining the existence of any lost data parcels so that retransmission of such pieces can be effected.

End-to-End Circuit Control -

This function serves the purpose of assuring that the transmitting point does not overwhelm the receiving point with too much data at once. This particular activity requires some coordination between sender and receiver, to the extent that the sender can send only certain groupings of sequences until permission is given to resume transmission of the next groupings.

Error Handling -

There are three situations that come under error-handling procedures: the positive acknowledgment, which is the event that precludes some indication of error; a time-out, after which retransmission is required; and a positive error detection that causes a request for retransmission.

Multiplexing -

This is the process of sharing and interleaving different data parcels that are using the same physical transport mechanism. Each data parcel's transmission facility had previously set up a logical path for routing data parcels, called a virtual circuit. Physically, however, this virtual circuit is in reality a transmission line that can manage the flow of many data parcel sources at once, using a higher bandwidth to orchestrate the volume. The process is reversed at the receiving end so that data parcels designated for a specified destination are routed locally to the correct termination point.

Figure 11-7 illustrates how a protocol operates. A message works its way down one chain of the communications layer stack and back up the other side to the ultimate destination. This process requires administrative control at each layer. At each level of administrative control, information is attached to a message or message part, and the message length changes. As the message travels back up the other side of the message protocol stack, this administrative control is stripped until the message reaches its destination. Peer-to-peer communication is accomplished indirectly by transfer from one applications layer interface down through one set of layers and back up through the corresponding set of layers to another layer of equal rank to its final destination.

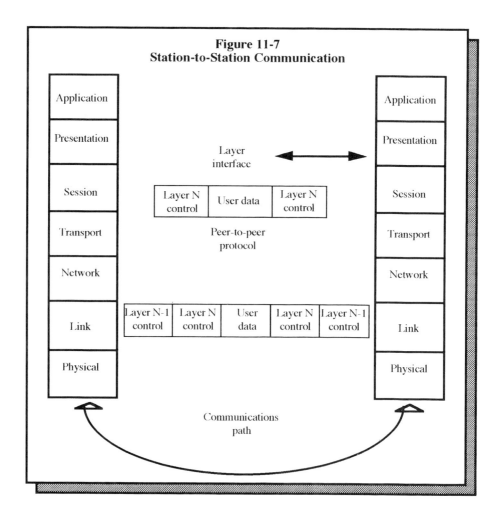

Figure 11-7
Station-to-Station Communication

11.3.1   Multimedia Network Issues

Multimedia conversational sessions involve what can be referred to as packet interactions. The newer broadband technologies, such as ATM, FDDI, frame relay, SONET, etc., have made available the platforms necessary to implement packet systems of divergent types and from divergent sources. Some of the older packet technologies, such as X.25 and TCP/IP, will continue to be utilized in these systems for many years into the future, but their utilization for such requirements as packet video and packet audio will be limited at best, because of the restricted bandwidth presented by these technologies. Figure 11-8 illustrates the conceptual usage of packet technologies for the multimedia integration of the principal media.

At least one platform is needed between communicators so that the different modalities can be passed either individually or as an integrated whole. The advantage of packet technology is that it makes no difference whether or which modality is being passed at any particular time. Identifiers within the messages flag these conditions and allow the receiving end to route the appropriate mode to the suitable hardware. This situation illustrates the key to multimedia network management. While the normal node and connectivity issues still occupy the usual network management implementation, the applications in the media network dictate the multimedia network management dilemma in that the quality of the network is more a matter of the quality of the applications programs than the data systems. Each of these technologies will be discussed below.

11.3.2   Network Principles

11.3.2.1   Connectivity

Connectivity may be represented by a multiple set of fields in the database for each equipment item. The connection linkages between equipment items should also be a part of the total picture provided by the configuration management subsystem. These links may be of several types, either dedicated or dial-up transmission circuits, whose characteristics can be reflected in the database as well. Relating both together, as indicated in Figure 11-9, is accomplished through cross-referencing equipment ports with specific communication links. The definition of circuits may take on such attributes as:

(1)   Identification number

(2)   Type of service

(3)   Bandwidth capacity

(4)   Quality

(5)   Attachments, such as equipment connections

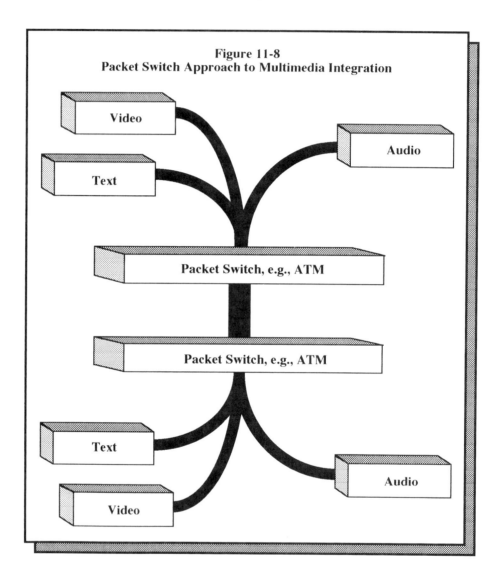

Figure 11-8
Packet Switch Approach to Multimedia Integration

11.3.2.2   Performance Management

Performance management is a means of tracking operational status and is
configured in much the same way as turnstiles that track spectator access to
sporting or entertainment events, i.e., the subsystem measures inputs and

outputs. Depending on the nature and sophistication of the system, performance management may take on additional tasks under the heading of planning support, such as the projection of loadings under various circumstances and the recommendation of actions to be taken as a result of such projections. The tracking of operational status itself has two aspects, one of which is the collection and refinement of statistics and the other of which is the use of these statistics in other analysis tools, such as queuing theory. This set of possibilities is organized in Figure 11-10.

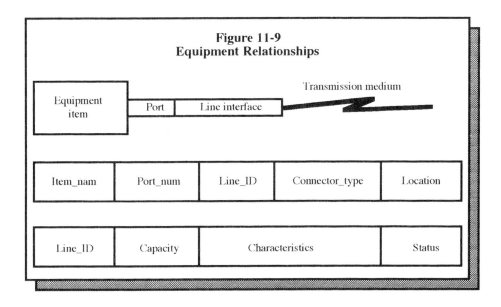

Figure 11-9
Equipment Relationships

The collection and analysis tasks require a database system that can assimilate data from many collection points and that can accommodate duplications. The design of such a database should ideally include:

(1) The type of data or data block being transmitted

(2) The data size

(3) The location of the collected status

(4) Source (origination)

(5) Destination

(6) Network type, e.g., X.25, frame relay, or TCP/IP

In essence, the performance subsystem should be capable of identifying all the envelope or external characteristics about each transmission within its grasp. Analysis will provide a statistical picture of the ensemble of events passing through a specific part of the system. In reality, however, not all envelope information may be available, and samplings of traffic may be all that is practical.

Loading projections and routing recommendations could be handled in the following manner. Trend curves and/or other historical reference algorithms could be used in the loading projections to make future indications of activity for each analyzed point. The paths to be studied might each be subjected to different curves and analysis algorithms for proper analytical results. Such complex analyses can be done most effectively by using computer systems with the necessary projection programs and databases. In many cases spread sheet-like tools suffice to track and work the projections under analysis. Such analyses as these can be used to plan new or expended services in specific route structures.

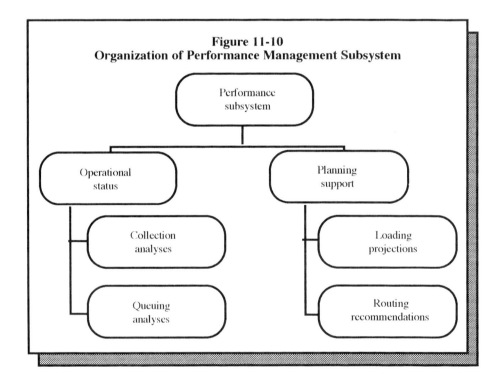

Figure 11-10
Organization of Performance Management Subsystem

Summary

In this chapter the various parameters that form the configurations possible under the multimedia network management constraints are presented. The key to the management of multimedia networks is their level of quality, but on a lower level, the management still depends upon a solid network management protocol. This chapter discussed just such management schemes.

References

(1)   Marney-Petix, V. C., *Networking and Data Communications*, Reston Publishing, 1986.

(2)   Stallings, William, *Handbook of Computer-Communications Standards*, Macmillian, 1987.

(3)   Tanenbaum, A. S., *Computer Networks*, Prentice-Hall, Englewood Cliffs, NJ, 1981.

(4)   Terplan, K., *Communication Networks Management*, Prentice-Hall, Englewood Cliffs, NJ, 1987.

(5)   Yager, T., *The Multimedia Production Handbook for the PC, Macintosh, and Amiga*, Academic Press, New York, 1993.

12

Multimedia Network Scenarios

Chapter Highlights:

The next step for multimedia is to expand beyond the CD-ROM, laser disk, and desktop computer, which only a few technical experts, media wizards, and hard-core computer enthusiasts are privileged to use. In order for multimedia systems to become the next wave of computer attention, they must be made available to, and put into the hands of, the average person and, more important, the average person must become acquainted with and comfortable with this form of expression. This chapter covers and discusses several systems that may be seen in commercial form in the near future, some of which have antecedents already in operation. The personal aspect of multimedia as introduced in Chapter 1 is the central ingredient of all of the products and services discussed in this chapter. The implementing features of these systems require not only an ability to conduct interactive conversations with the user, but also incorporate intelligent systems to give functionality to the systems that will be designed and produced.

12.1 Promise of the Future

Up to the present, multimedia has been relegated to an implementation where the majority of the information used to support interactive capabilities is created off-line, that is prior to actual need, and then stored onto some medium, such as a laser disk, CD-ROM, or other high-capacity storage device. The information is then made available to the user by means of a local direct access scheme, such as by the local attachment to the using computer from a CD-ROM via its controller. The other major approach to the provisioning of a multimedia offering is the distribution of the information to the eventual user via some sort of on-line information service. On-line information services are sometimes referred to as remote access systems. Several on-line information services exist today under such names as *America On-Line, CompuServe, Prodigy*, etc. *Internet* may also qualify as a multimedia information service, depending upon the nature of the information being transferred between server and user. The nature of Internet is that it can be accessed through any of the on-line services currently available, such as through *America On-Line*.

Local or direct access systems consist of the sale, or physical distribution, of the medium to the eventual users, at which point the information can be accessed as needed. At some point, the information may need to be updated, and the process begins all over again using physical distribution as the primary means of updating the information [2].

Current on-line information systems have the distinct drawback of lacking the bandwidth necessary to sustain full-motion video presentations because the delivery and updating process is made via limited-bandwidth transmission capabilities, thus violating maximum delay requirements needed to support video and audio. This limited bandwidth condition will corrupt or prevent display delivery protocols from operating properly, thus distorting the delivery process.

There are some efforts being made to develop and make available a widespread set of facilities for digital and analog transmissions systems that can accommodate full-motion video over a host of media. These media systems include such technologies as twisted-pair copper lines, fiber optics, coaxial cable, wireless media, and other transmission systems that are widely or potentially available today. The general expectation is that still or stop-motion imagery will probably be the common method of conveying the image part of on-line services for some time to come.

Another aspect of the promise of the future is that we are moving rapidly into an age of the personalization of our purchasing, information needs, privacy issues, such as health information, and other topics related directly to our private identities. One of the primary motivations for multimedia has been and will continue to be the personalization issue. The exchange and use of information is primarily a personal concern, where the requester is motivated by specific questions.

Beginning now and extending well into the indefinite future, multimedia capabilities and systems will be of a different type. It is expected that the population will become more and more mobile. Additionally, financial transactions will, more and more, be void of actual money exchanges, and information will evolve into personal repositories so that each person can have access to his/her own personal databases. These databases may be known under other names than databases, such as directories and files, but the effect will be the same.

Multimedia will become a tool for those on the go [3]. Transportation hubs, living accommodation crossroads, and all mobile devices and mechanisms will likely be the scenes for interactive communications between individuals using all of the modalities available. The crucial element in the development cycle of the multimedia systems that are to be used in these environments is the structural design and designed-in utility of these systems. These systems can be categorized so as to define just a few types of such devices as indicated below.

(1)   Personal Multimedia Unit:

Personal Video Production -

This is a facility to create and distribute video generated by private individuals for the exclusive use of friends, acquaintances, and family, or to be used for business. The storage is accomplished locally and the transmission system is twisted copper wire. The reproduction of the imagery is performed at the receiving end using standard TVs or computer displays.

Personal Directory -

Seventy-five percent of the population in the U.S. and Canada relies upon his/her own personal directory for telephone numbers and ancillary information about these numbers. The only time when people use the printed directories is when the anticipated contact is new to the user. Contrary to what the Yellow Pages companies would like you to believe, commercial contact is generally made through the personal directory. If there were some way that the personal directory could be automated and periodically updated, its value would be increased in the eyes of the owner of that directory. A personal

directory is each person's own list of phone numbers
and addresses, coupled with incidental notes about
these references.

Personal Communicator -

This type of device is similar in objective to those
devices already appearing on the market today. Such
devices as the Newton Pad and others are attempts to
provide personal communications of the form most
familiar to consumers today, i.e., in hand-written form.
The device envisioned as the multimedia equivalent of
tomorrow's capability is a unit that not only has
communications capabilities, but that also
incorporates shopping features, and other more exotic
features, such as interfaces computer aided
manufacturing (CIM) services. CIM is a feature that
incorporates the owner's dimensions and expressed
buying needs so that the buyer's replacement needs
can be satisfied. Similarly, the personal communicator
also has a range of the owner's needs built in or
programmed in so that these can be addressed while
the owner is occupied with other issues.

(2)   Commercial Multimedia Devices:

Interactive Financial/Banking Unit -

The familiar automated teller machine (ATM) will
evolve to the point of becoming a self-contained
"bank in a box." Such a unit will have facilities to
communicate visually and verbally with a bank officer
as required to meet customer banking needs, such as
loans, refinancing, investments, etc. The unit will also
be capable of interactive access to the customer
accounts, transfer from account to account, access to
reference information about the local economy, public
finances, development of projections for the user, etc.
Finally, the unit will allow the customer to perform all
banking functions normally allowed in today's walk-
in branch banking office.

Public Data Retrieval Station -

The on-the-go consumer will have additional
opportunities to obtain business and personal
information while passing through airports or other
transportation facilities. These units will also be
located at hotels and other places frequented by the
busy traveler. The facilities to be implemented will
include capabilities that would ordinarily not be
available to the traveler with a notebook computer.
The additional features envisioned include the ability
not only to access remote databases and other
repositories, but also to retrieve information of interest
and to commit that information to nonstandard
storage media, such as VHS, Video 8, cassette.

Home Services Unit -

The home services capability as expressed by the unit
designed for that purpose is essentially a home
shopping facility that provides the consumer with the
ability to do on-line comparison shopping, locate
hard-to-find items, and select the consumer's choices,
all from a standard unit. This unit will be driven by
information obtained from remote databases via
telephone connections.

Government and Civil Affairs Booth -

Certain civil and governmental-voter interactions
can be effectively implemented by a series of
methods, among which are home computers and
public kiosks. For voter and/or citizen surveys,
home computers using modem interfaces with data
collection computers may be sufficient for survey
validity, but actual electoral registration, i.e., voting,
may be a different matter, as voter authentication
may be the only method to validate the results. As a
result, kiosks located in a public place may be the
only acceptable method of collecting results. Such
devices may also be used by the voter for accessing
information about choices and looking up historical
reference information. The unit will have the ability

to access remote databases, to view animated clips, display video segments, etc.

Health and Human Services Kiosk -

An example of many other types of special-purpose facilities that may be used for specific purposes in the multimedia arena is the health and human services kiosk. This device, like others, such as a medical advice, or tax preparation unit, has capabilities that are suited for the purposes for which it is designed. This device might be utilized for the access of job opportunities, for programmed instruction to train people for certain job positions, to provide computer-animated scenarios for other job tasks that may be required for certain training positions, etc.

Tourism Planning Kiosk -

This unit will probably be used extensively during the next few years. Versions of it are already in trial usage in order to provide proof-of-concept. The tourism kiosk will allow the tourist to plan trips, purchase tickets, determine services, print desired information, see video clips of items of interest, and the like. The locations for these units will either be close to the attractions of interest or in areas where the travel is thought to commence.

These are some of the basic types of systems with which the citizen of tomorrow will interact with his/her private and public life. Not all of these systems will be suitable nor even available for private use in the home, on the one hand, or on the road, on the other hand, but their need and use will be ubiquitous, whether in the shopping mall, airport, hotel room, automobile, or home. The basic features of all of these units will probably also be generally the same. The anticipation is that each will incorporate the following general features.

(1)   An extremely easy interface, or method of interacting with the multimedia device

(2)   A communications interface with other users, or the database repositories with which the multimedia device is communicating

(3)   Applications facilities for generating the desired mixture of voice, print, and video components

Discussed below are some of the multimedia implementation devices that may be introduced into the commercial marketplace within the next 2 to 5 years [2]. Because economics is the great motivator, the features, functions, and delivery of any of these services will be dictated by some sort of cost-effective formula developed and exploited by the providers of such facilities. Also, the network management features and essentials of such multimedia systems will be of prime importance in terms of meeting user expectations. Both of these issues will be covered as each example is outlined below, along with the basic service description of each system.

12.2 Multimedia Banking Terminal

The automated teller machine (ATM) has gained widespread acceptance in the personal banking and financial markets over the past 10 years. The basic ATM represents much more than just a high-tech curiosity. It is one of the first instances of mass audience acceptance of a high-technology interface, following the mechanical parking meter. ATM devices are further advanced than the parking meter in that they are interactive in the sense that they require the user to enter certain information in order for the machines to deliver the desired services, such as account status, money, funds transfers, deposits, etc. In this section, we will investigate the next evolutionary step in the personalization of our daily lives as facilitated by an advanced ATM.

12.2.1   Application Overview

The use of ATMs is still increasing in North America at a double-digit rate, even after experiencing a sustained rapid growth rate in the 1980s. These units have several consumable attributes, among which are that they are a repository of currency, ink, print paper, and envelopes used for deposits. All of their other attributes are electronic in nature, and therefore require only the standard input devices and a display monitor in order for the everyday user to take advantage of their programmed features. The time has come when the ubiquitous ATM is ready for a new generation of capabilities which will make this device more able to support all banking requirements of the average depositor. The timing for this evolutionary change is due to several technology advances among which are data and video compression, applications programming, data security, and the like. Several of the more important generic features of this advanced ATM (AATM) are as follows.

(1)   The use of ATMs will incorporate the exercise of a 2-way video capability to provide personal interaction between customers and banking agents. The bandwidth associated with any 2-way video transmission is scarce, and therefore various coding and compression techniques will have to be used in order to achieve the aims of this objective.

(2)   The unit must also be equipped with FAX and/or scanning capabilities to provide written document interchange. Legal and official document exchange is crucial to the utility of the AATM. Documents can be scanned for viewing by the central facility personnel having video conversations with the customer, and each AATM will also have a depository so that the documents scanned can also be captured for legal proof of usage.

(3)   The unit will provide most, if not all, banking tasks, eliminating or displacing the normal "walk-in" branch bank operation so familiar to the average customer. This set of features will favor greater convenience and more centralized control for bank operations, resulting in operational savings.

(4)   Typical business transactions might be loan applications, checking account questions, funds transfers, etc. Each has its own unique set of business parameters with which it must operate, and the media requirements are as varied as the parameters involved.

(5)   Document scanning will be required so as to facilitate the exchange of official printed matter. The differences between FAX and document scanners is one of conveying a document already in hard-copy form that is to be received also in hard-copy form, as with a FAX capability, and a hard-copy form that is intended to be transmitted and received in electronic form, such as in a document database format.

12.2.1.1   Market Opportunities

The idea behind the advanced ATM is to target branch banking operations. Current bank branching activities are manual, very labor-intensive, cost-inefficient, and scattered from a management oversight point of view.

The demand for access to financial and banking facilities is growing in spite of banking industry consolidations and down-sizing activities by various banking companies. There are currently about 100,000 ATMs in operation nationwide in the U.S., about 10,000 in Canada. Additionally, there are 46,850 branch banks in the U.S. and about one-tenth that number in Canada. The industry trends are toward the reduction of bank branches for cost containment reasons, but the ATM cannot absorb the new functionality requirements of the branch bank until something is done to upgrade the basic ATM. Thus, an advanced version of the ATM in a multimedia envelope is required in order to accommodate these new market needs and cost requirements.

The number of ATM transactions per day in the U.S. is over 16 million. The number in Canada is about 10 percent of this value. This represents only about 25 percent of the load that could be handled by existing systems in the U.S. if the maximum loading of these systems were based upon a 24-hour-a-day operation. This is not a very realistic possibility, but this value does indicate that there is a lot of room for communications expansion. Because new systems would have to be equipped and installed to support this new concept in remote banking, as described above, the timing for the introduction of the advanced ATM has to be done at a point where the economics are right.

The bottom line for the advanced ATM concept is that, on the one hand, the current generation of ATMs has not yet reached saturation level. On the other hand, the economic pressures to reduce operating costs have created the need for something to be done to continue servicing customers while at the same time cutting back on the number of branch banks in the U.S. and Canada.

12.2.1.2   Key Features and Capabilities

If the tactics of banking houses, are, as they seem to be, to provide greater access and availability to specific banking services, and greater personalization between the customer and the bank, then the system that can provide this set of features is one in which the access is made as geography-independent as possible, and one in which the personal touch is made through face-to-face interaction, even if via a teleconferencing-type system.

The capabilities of such a system will have to encompass all of the present-day services, such as cash access and account queries, but also will have to provide face-to-face interviews between banker and client. As with current banking operations that include phone, ATM, and walk-in access to the banking facility, this new system will also have to provide security and privacy for the bank as well as the customer.

The capabilities envisioned for this new-generation capability will be expanded ATMs equipped with 2-way viewing, FAX, scanning, voice, display, and query facilities. These units will reside in localities independent of current branch bank facilities, such as in malls, public buildings, and so on. The capabilities would also be enhanced by the inclusion of video interactions that provide personal contact. Figure 12-1 illustrates the overall architectural concept for this capability. There is a communications interface that sends and receives the various modes of communications products, which, in turn, is supported by a multiplexer. The human interface is elaborate and complete to the extent that it will support all banking needs of the consumer. There is 2-way image exchange for document transfers, and 2-way teleconferencing for human interaction with other humans.

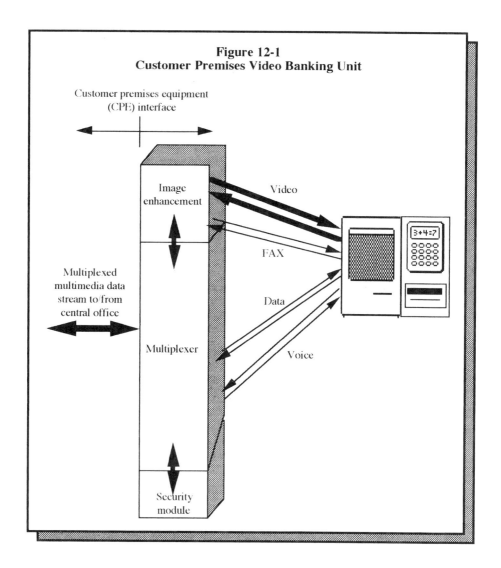

Figure 12-1
Customer Premises Video Banking Unit

12.2.1.3   Market View

The growth rate for ATM transactions in the U.S. is 12.2 percent compound annual growth rate (CAGR). Again, for Canada the growth is about the same. Because of this and because of the transaction numbers that are already in evidence, it is clear that the time for an advanced version of the ATM is near at hand. The actual transaction numbers of 90,000 plus indicate that the ATM has gained widespread acceptance across both the U.S. and Canada, with the actual number of branch banks in the U.S. and Canada being about 52,000. If the operating costs of each branch represent about $700,000 per year, and if the operating costs of an AATM are $100,000 per year, well over a half-million dollars per annum could be saved by converting to the AATM concept, even if 2 AATMs are located at each site where a branch was previously located.

The number of daily ATM transactions has grown rapidly over the past 10 years, to the point where the average daily number of ATM transactions in the U.S. and Canada is over 22 million. This represents over 253 transactions every second.

Thus, the advanced ATM has a market entry position that can be justified based upon the large volume of transactions already performed daily in the U.S. and Canada. Finally, the low operations costs of the AATM and the growth rates of ATM usage all point toward a situation where an interactive system similar to the present ATM booth is ready for introduction into the banking marketplace.

12.2.1.4   Network Services

The architecture of the advanced ATM (AATM) system is based upon 2 platforms, one being the AATM placed at various convenient locations where customer needs demand, and the other being the AATM at the centralized, or district locale. Both of these node types will be discussed below. The network service(s) needed to accommodate these features from a transmission network point of view is outlined as follows.

(1)   A bandwidth to supply 2-way video between booth and bank facility, 2-way digital link for normal display and usage, 2-way FAX link, and 2-way voice link.

(2)   The network services needed may also include data compression and security features, e.g., authentication of location, data encryption, etc.

(3)   Service would most likely be multiplexed and, most likely, packet-switched. The delivery of the various multimedia services will require multiplexing at the central facility as well as the branch locations.

12.2.1.5   Functional Alternatives

The alternatives to this type of approach to cost reduction and customer satisfaction needs are rather limited given the situation facing the banking industry. First of all, the banks could simply contract their banking branches by closing selected branches, and let the customer base decide whether this condition meets their needs. Second, the bank could encourage its customers to become more involved in bank-at-home options. This approach has several attractive features, because the home computer has many capabilities that a computer system did not have just 2 years ago. The problem of obtaining cash is still an issue that must be accommodated by the bank. A third alternative involves the creation of an advanced ATM which has all of the capabilities that are needed to service the customer's needs. These alternatives can be summarized as follows.

(1)   Force customers to use a decreasing number of branch banks, with the attendant risks of customer dissatisfaction and defections to other banks.

(2)   Encourage customers to use mail more often, as a partial alternative.

(3)   Some banks may consider repositioning bank business away from the consumer market toward the business market.

(4)   Develop and market alternative technology solutions, such as bank-at-home systems, using video boards on home computers.

12.2.1.6   Technical Issues

The multimedia implementation of interactive banking systems requires a definition of network services and the customer interface device in order for such a capability to be realized. As discussed above, the network services required include 2-way video, FAX, audio, and data, all of which are multiplexed for relay to the central processing facility of the banking concern, and which will be demultiplexed at the receiving end where the multimedia booth is located.

The basic technical components of the banking data booth include the human interface, the communications multiplexer, drivers, etc., the image processing capability that may include compression/decompression algorithms, image enhancement facilities, and quality measurement features.

12.2.2   Network Management Issues

The network management of the banking data booth requires the same approach as the multimedia concepts already introduced in this text. Quality is the most critical issue facing the user of the services provided by the system. As a result, the management of the service will emphasize the physical layer and the applications layer. The network manager will monitor both the multimedia interface and the user interface equipment as well. The network management concept is illustrated in Figure 12-2. In this figure, the 3 principal elements of the system are: (1) the communications interface; (2) the booth or kiosk; and, (3) the human interface facilities. The kiosk capability is provided by a combination of hardware and software capabilities, notably teleconferencing, document scanning, text-to-speech, video, and document retrieval. The operational concept is that the customer will enter the booth, provide some type of identification, such as retinal image, fingerprint scan, etc., and then conduct his/her business.

The manager must be capable of accessing all AATMs from each of their diverse locations. In order to accomplish this, these units must have either an in-band or out-of-band signaling (communications) link between the manager and each of the AATMs. In-band signaling means that the network monitoring and control is handled by the same circuit as that which handles the multimedia interchanges. Out-of-band signaling indicates that the network management signaling is handled by a circuit separate from those exchanging the operational data. There are advantages and disadvantages to both approaches, but these are a function of the importance, finances, and practicality of every situation, so there is no right answer.

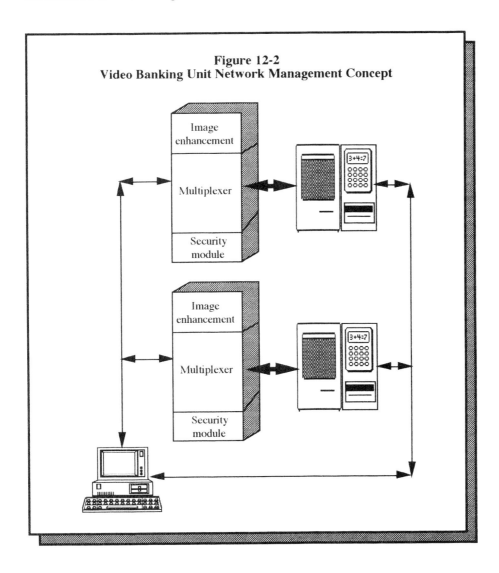

Figure 12-2
Video Banking Unit Network Management Concept

The network manager will have to be aware of the activities within each session so that if an outage occurs, the manager will be able to pinpoint what was happening at that time for purposes of understanding what went wrong. This is necessary because, for instance, a scanned document transmission may be in process which is to be used for official purposes, and therefore key to an interactive transaction. The physical linkage between each AATM booth and the central facility is the first line of defense in detecting faults on this system, while quality measures can provide information about the nature of the document transferred between customer and customer service personnel.

The super booth attempts to make such "walk-in" situations more regular in terms of their arrival. Queuing theory teaches that regular customer arrivals versus random customer arrivals for a single server queuing system yield a queue length of about half the number of customers expected with the latter queuing discipline. If the staffing picture is used as a criterion, the staff to service the arriving customers could be reduced by one-half to achieve the same results as when the customers were bunching their arrivals. A 50 percent cut in special customer service staff, e.g., those providing investment and loan support, could be significant to the information service management.

12.3 The Personal Information Booth

Another multimedia device and assembly of system components that will probably be available for the use of the everyday consumer, especially at travel crossroads in the near future, is the personal information booth. This unit will provide a particular segment of society with a method of sending and retrieving information of a personal or even a business nature. The major features of this system are the ability of the system to capture, retrieve, and store onto various media the information needed by the individual user. The tools include those needed to deal with multimedia formats of all types, as well as the ability to display high-resolution imagery and graphics.

The commercialization of such a system will be realized as the potential market and the providers come to the conclusion that this is a need that must be filled in a way that is understandable by the target market. The question is: What is the nature of that interface?

12.3.1   Application Overview

The potential for the data booth or any other multimedia system is tremendous if it catches on with the market to which it is targeted. Whether or not it is a success depends upon its functionality and design. Some of the more important aspects of this opportunity are discussed below.

The first multimedia device and system configuration was showcased as part of a laboratory demonstration that occurred in the midtown New York laboratories of AT&T in the 1920s. Bell Labs developed 2 such units and demonstrated their potential between 2 of the Bell Labs facilities in the Manhattan area. The demonstration incorporated 2-way TV and audio, so that each participant could see and hear the other party during the conversation. While the operators were viewing each other, one operator would vocalize a thought that would be received and heard by the other operator. They could control their cameras so as to pan them about the environment surrounding each operator.

The video phone concept, first unveiled at the World's Fair in 1960, was the first instance of a multimedia system that had reached a point of concept demonstration. With some additional work, the idea could have reached a performing field trail within a few years, but it was held off for several reasons, among which had to do with the bandwidth required to support the full capability of the operation.

The data booth is an outgrowth of these various attempts to provide natural conversation between humans even though they are many miles apart. It also has its roots in several attempts to provide the consuming market with data access, a need that began growing rapidly in the late 1980s. Many ideas, centering around the notion of the public kiosk, were investigated, and some were even trial marketed. These systems targeted public information, travel/tourism issues, and public event ticketing. During the 1980s, display phones began appearing in airports that were teletext-oriented, providing usage information and other services that the provider was tempted to offer. These devices are also fully capable of providing other display offerings, such as graphics, imagery, and full-motion video. At some point, the providers of these installed units will define and implement a set of capabilities that may be the beginnings of a fully commercial data booth.

Another small instance of multimedia interaction can be claimed by the cable TV industry with its pay-per-view (PPV) innovation. The architecture for this capability is implemented in one of 2 primary ways, one being the use of the telephone to specify and validate the choices made, and the other being the use of the CATV system itself to provide upstream signaling back to the headend with the choices selected. In either situation, the viewer signals the headend as to its choices, after which the system responds with a media presentation.

As the years progressed, bandwidth problems have been attacked and solved through research in and development of coding and compression schemes that increased the effective bandwidth of video data streams by utilizing digital formats rather than the analog format that had been used exclusively in the first demonstrations of multimedia in the 1920s. In addition, digital formats have facilitated the implementation of 2-way signaling as well as delivery of the media signals by allowing the segmentation of the bandwidth for each of the forward and backward signal paths.

Figure 12-3 shows a front view of the data booth layout. The unit has several input/output facilities that are the key distinguishing characteristics of the system itself. In addition, the ubiquitous telephone handset is provided, along with a display screen, keyboard, touch-tone keypad, and a slot for a receipt or printer feed. Outwardly, the unit is not very different from the AT&T

display phone found so often in airports, except for the variety of tape devices in evidence in the figure.   Such a unit is ideal for FAX transmission and reception, retrieving voice messages, text-to-speech conversion, voice-to-text conversion and storage, transmission and retrieval of documents, creation of presentations, etc. The various input and output (I/O) interfaces are necessary in order to satisfy the many divergent needs for such I/O.

12.3.1.1   Market Opportunities

The viability of a data booth for transient traffic usage, such as airport locations, is still undefined at this time. The assumed requirements are that the system have the following capabilities:

(1)   Multi-formatted input-output (I/O) so that the user can send or retrieve computer disk, Video 8, VHS, or cassette-formatted materials.

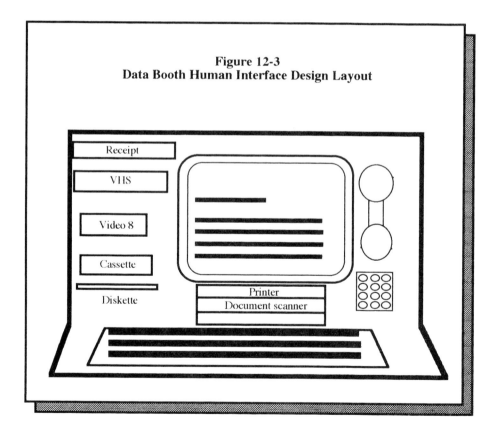

**Figure 12-3
Data Booth Human Interface Design Layout**

(2)   Access to on-line information systems, such as *CompuServe*,
       America On-Line, *Prodigy*, and other disk media.

(3)   Software to transfer accessed information between their
       repositories and the storage media available at the data booth.

(4)   On-line capability to create, view, edit, and otherwise manipulate
       the data brought in from outside sources or input from ports on
       the unit. This means that the software has to have presentation-
       level protocol interpretation and conversion capabilities.

The basic usage scenario is for the user to sit down at the unit, provide
some sort of authorization, such as a credit card, insert his/her storage medium,
retrieve the source data, review and change the data as required, and store it on
the desired medium. The types of information of interest may be e-mail,
presentation material, animation, documents, and the like.

The customer base needs to have a special interest and technical
awareness of the capabilities to be marketed. This defines a special customer
segment, which may narrow the market considerably.

12.3.1.2   Key Features and Capabilities

If the functions and ancillary capabilities were marketable, the market segment
to which they were targeted would have to be identified and then the financial
values would have to be assessed as well. A reasonable guess would be that
each such session access would be valued at between $2.00 and $5.00.
Additional costs would be assessed for extended time usage for individuals on
the go, such as sales, professionals in various fields, and technical support
personnel. The key to this and the other forms of multimedia usage terminals is
that they must be capable of tasks that cannot be easily expropriated by
notebook computers. In this scenario, the one element that is beyond the
present capabilities of the compact computer system is the array of storage
media inputs. Coupled with this is the number of algorithms that must
accompany such a variety of storage media to read and convert between the
various media, and to convert between the various presentation-level protocols.

12.3.1.3   Market View

The market available for this set of features can be described as having 3
components, total market, addressable market, and target market. The total
market is all the people who use some form of data booth facilities in the
territory of interest. The addressable market includes those potential customers
who use the various means of information and data capture, from FAXs to
Federal Express services. The target market is the total of customers who could
use such a service in a transient travel area. Thus, the market for this

application includes addressable market customers as well as target market customers.

The window, or time frame, for this type of capability is hard to assess, because there are no current foreseeable constraints that would limit the time frame for such a new product introduction. The size of the market, however, is a little more definable. If we assume that the market is roughly 5 percent for early adopters and another 5 percent for that population segment that is attracted to new technology, then about 10 percent of the target market would likely be customers for this system, at least initially.

12.3.1.4   Functional Alternatives

Outside capabilities that could displace the features and capabilities proposed here to some extent include video phones and the regular video-teleconferencing systems available today. Neither of these has the interactive features envisioned by the development and employment of the data booth, nor is their utility for such an application clear at this point, either operationally or fiscally. Notebook computers, on the other hand, have the convenience, size, and full range of capabilities needed to satisfy most communications and computational needs today, with the possible exception of the different media interfaces that may be demanded by the prospective user.

12.3.1.5   Technical Issues

The basic technical approach is to envision a multimedia "phone booth" that has the features identified above. The video capability is provided through a switched network service of fractional T-1, full T-1, or wider, bandwidth. The capability will have to be 2-way so that customer and banker can each see the other simultaneously and talk to each other as well. This capability will be provided via switched services to accommodate ubiquity of location access in the area of interest. The bandwidth will be able to accommodate the following information streams: (1) 2-way video; (2) 2-way voice; (3) 2-way FAX; and, (4) 2-way keyboard and display interaction between the customer and the information repository computer.

Operational approaches to this network service include the following considerations:

(1)   Attractiveness of defining the service so it can be tariffed as a unique and identifiable offering to business

(2)   Providing this service as a switched feature of the network

(3)   Providing separable features that can be identified as network-based capabilities

If these data booths can be thought of as essentially super phone booths, their operation will require connectivity to the network, multiplexed voice, print, and video capabilities, dial up capability for linkage to any of several different information services facilities, and several possible special features that would provide acceptable video. These special features might include network-based coding/decoding hardware as well as image enhancement facilities.

The following summarizes the technical efforts, known thus far, to be required:

(1)   Demonstration of a dial up T-1 or fractional T-1 capability in the areas to be tested or designated for installation

(2)   Demonstration of frequency spectrum allocation for simultaneous 2-way video transmissions, 2-way voice, 1-way print, and 2-way FAX (possibly half-duplex)

(3)   Demonstration of acceptable video presentation for compressed bandwidth transmissions

(4)   Demonstration of network ability to "trap" duration of transmissions made during customer use of the super booth, as well as the nature of these transmissions, e.g., FAX transmissions, message print, etc.

12.3.1.6   Development Issues

From a system development point of view, there are several components to the service to be offered. These include the development and system integration of the facilities (super phone booths), the definition of the services to be offered, the market analysis of the opportunity, trials of the service with real clients testing real customers, and possible advertising possibilities that could defray expenses of operation, that would be displayed on the video screen prior to a transaction.

It appears that standards have already been developed that might be used in the implementation of the offering, such as those that establish the transmission facilities to be required. Given this, the implementation approach would be to survey the users' data needs, and to provide the developer with preliminary marketing analyses and a concept description. The operational concept would include privately owned and maintained data booths, with multiple access to any of a number of information services providers that may subscribe to the service as a provider. Service costs would be chargeable on a pro rata basis depending on usage loading or a flat fee for a certain time limit or a combination of both. Next, based upon interest, several booths would be built, dispersed at different locations, tested for utilization and labor hour savings, ease of use by the public, and customer acceptance. The booths would be equipped and integrated using off-the-shelf technology for display and transaction interfaces. The network side of the operational implementation would be provided by coding/decoding (codec) equipment currently available to compress the video data. The actual manufacture of the booths and the installation of equipment would be contracted to a third party.

The stakeholders would include the providers of protocols for information services access, applications access, format conversion, manufacturers, potential customers, competitors, and the like. These various interested parties are discussed briefly below.

Manufacturers -

Companies that produce kiosks and other information-type booths, such as ticket dispensers, self-service FAX machines, etc., may be especially interested in furnishing these types of devices for public usage.

Customers -

As a group, this target market might resist the phase-in of this new service, having just gotten used to the ability to use the notebook computer with all of its capabilities and usage procedures.

Developers -

This group is likely to be very supportive of this new system for obvious reasons, including the creation of intellectual property, fees for development services, etc.

Telecomm Carriers -

The long-haul carriers may argue that they have a better chance of making such a system work because they have the charter to provide long-distance services. Since much of the transport for these services could be long distance between the booths and the centralized provider facilities, there may be strong support in favor of this argument.

Information Services -

Aside from the equipment manufacturers, information service providers may be the biggest supporters of this new system. The support of these providers is key to the acceptance of this system.

12.3.2   Network Management Issues

This type of network is not too dissimilar from the advanced ATM as discussed in the previous section. The network architecture concept is shown in Figure 12-4. This type of system has units that are geographically dispersed within some area of interest. Additionally, there is a network control center also located somewhere within this area. Each of the personal data booths is connected to the network manager via either direct or indirect interfaces, i.e., in-band or out-of-band signaling. The functional requirements of the architecture dictates that, at a minimum, the following functions be monitored and evaluated as needed.

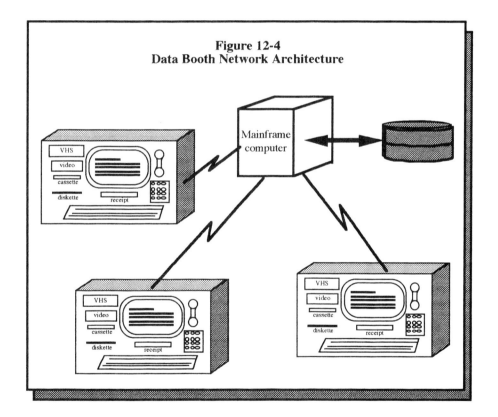

Figure 12-4
Data Booth Network Architecture

(1)   Each component of the personal information booth, such as the communications interface, human interface, processing subsystem, etc., should have its own outlet to the network manager for purposes of identifying its own unique operational status at any particular point in time. These monitored points will yield information about the health and status of each of these subunits.

(2)   Operational data about the actual information interchanges is important in order to assess, when necessary, what the status of exchanges is. These status items help to determine where the users are when certain outages, errors, or below-standard-quality exchanges occur.

(3)   The communications between network elements is the last major feature of the network control picture that must be monitored and periodically evaluated in order for the network management process to be complete.

One of the key aspects of the this type of system operation is that it invariably involves interactions between one of the personal information booths and a database system, whereas the advanced ATM involves information exchanges between both databases and humans, and between humans and humans via teleconferencing links.

12.4 Movies-on-Demand Paradigm

The delivery of electronic entertainment is a potentially profitable business activity. Among the various types of entertainment available are movies. Currently, movies are carried and delivered to viewers by such means as the familiar CATV network, both wired and wireless, movie theaters, over-the-air broadcasting, video stores that rent or sell these products, etc. Some of the various CATV networks have special channels set aside to deliver certain movies, similar in style to the in-room movies that may be purchased in hotels and motels. Such offerings are referred to as pay-per-view. These offerings are requested by 2 principal methods; one is the use of the common telephone to set up the request to either a person or a computer, and the other major request method is to request the movie directly through the CATV system using a subchannel on one of the delivery channels.

There are many variations on the theme of movies-on-demand. The most familiar is that illustrated above, while others exist as well. One of these is the use of special-purpose transmission systems, such as wireless or twisted copper wires to convey the movie signals. In addition, the request signals, sometimes referred to as upstream signaling, must be accommodated somehow. This can also be accomplished by using a small segment of the total bandwidth, called a subchannel, to provide the necessary upstream signaling.

The regional telephone companies in the U.S. have been trying to find ways of launching, both legally and technically, video services since the divestiture of AT&T in 1986. One of the popular visions of movies-on-demand envisioned by the regional telephone companies is the usage of the common copper wire infrastructure to carry such movie deliveries. This approach is illustrated in Figure 12-5. This figure shows the use of the telephone network to service the video request on the upstream side, i.e., from customer to provider, while the delivery is accommodated on the downstream side, i.e., from provider to customer, by a service bureau acting as the headend to transmit the video signal. The telephone switch, known as a digital access cross-connect switch (DACS), receives the channel request, sends it to the service bureau, and then switches the desired video stream out to the requester. The output may be routed either back through the switch or directly, depending upon the type of capabilities available within the switching system.

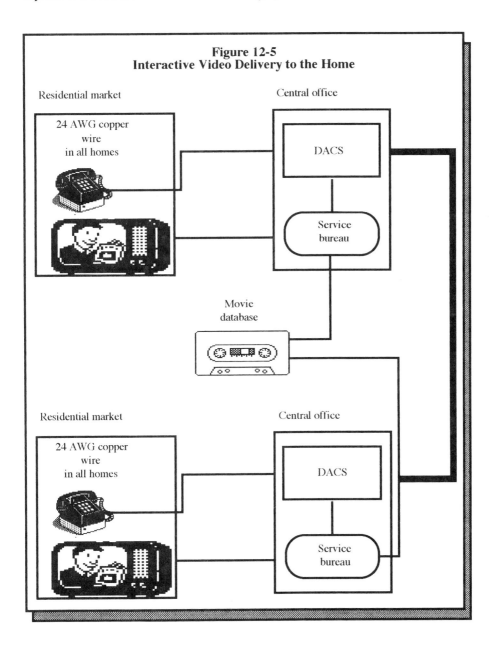

**Figure 12-5
Interactive Video Delivery to the Home**

In the common CATV, cable TV, approach, there is a distribution center, known as the headend, that provides all of the channels available to the viewing customers. The viewers choose any of the selections (channels) available from the TV set by tuning to the appropriate frequency filter. In the copper distribution approach of the telcos, the limited bandwidth of the copper telephone system requires that selection be made from the headend and switched to the user as opposed to selection at the display set.

In one design approach, the video signals are digitized and compressed prior to transmission. The major advantages assumed by this approach are the preservation of signal quality and the ability to transmit the signal at an acceptable signal-to-noise level over a considerable distance because of its digital nature, which is easily preserved by periodic amplification. Another approach is to retain the analog format of the original signal and to transmit it over a limited distance to another analog receiver. The assumed advantages of this approach are the reduced costs associated with maintaining the analog format. Another major advantage is the simplicity of the overall architecture involved. A final advantage might be its carriage distance even though the signal is analog rather than an equivalent digital form.

12.4.1   Application Overview

The market for entertainment material, such as movies and the like, is tremendous. Because of this, the need for distribution systems to carry and convey these types of services is significant, and distribution systems are also an area of intense research and development. All major distribution industries, such as telephone, CATV, direct broadcast satellite, video stores, etc., have their own ongoing video service delivery programs in progress.

The operational concept for the conveyance of movies depends upon the architecture under consideration, but the general customer needs are:

(1)   Wide variety of choice

(2)   Efficient selection of choice

(3)   Efficient delivery

Aside from these needs, the exact technology used for such a service is wide open for discovery. The telco distribution approach revolves mainly around the use of copper for the distribution of the signals of interest. This service has the following characteristics.

(1)   The movies reside upon either a disk or in a video tape library and are periodically withdrawn and inserted into a studio-quality VHS

player. The VHS tape players or the high-capacity disk equipment, plus the modulation and amplification equipment, constitute the headend or origination system that distributes the information.

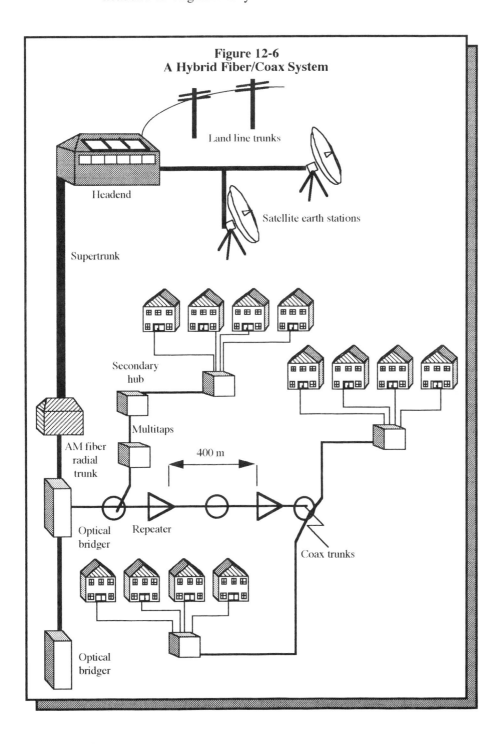

**Figure 12-6
A Hybrid Fiber/Coax System**

(2)   The headend is tied in to the distribution system at this point, similar
        to a telephone switch interface with lines that enter businesses and
        homes. One form of distribution system is similar to that shown in
        Figure 12-6.

        In this figure, the distribution is accomplished by a combined
        fiber optic and coaxial cable system that carries the demand signals
        from the headend to the destination point. The request portion of
        the loop can be realized by any of several methods, such as the
        telephone as illustrated in Figure 12-5. Thus, the solution to the
        problem of program requests can be satisfied by the use of the
        telephone in all cases, including copper architecture, coax systems,
        or even wireless solutions.

(3)   The other major method of request signaling is by upstream
        signaling from a set-top box to the headend unit. The major
        drawback for this approach is that coaxial systems use amplifiers
        about every 400 m to boost the signal, due to the extreme signal
        decay that occurs through such transmission media. The noise that
        is generated during these periodic signal boosts is additive, so when
        a message is transmitted in the opposite direction, it is facing a very
        high noise level. This problem has been partially remedied by
        installing 2-way repeaters in these branches of coaxial cable
        transmission systems. However, the problem has not disappeared
        entirely.

(4)   One of the methods being promoted by the telephone companies
        in North America is for video to be transmitted by copper wire.
        Under this proposal, the request will be sent upstream from the
        requester to the video headend equipment. At this point the
        requester's account will be accessed to determine if it is in order,
        and, if so, the request will be honored. The method for honoring the
        request will be by inserting the video signal into a portion of the
        total bandwidth possible for the copper transmitter and the distance
        over which the signal is to be sent.

Figure 12-7 illustrates the process involved. The request is handled by
the headend equipment, and the video in question is played into the network
with the destination address of the requester location. The signal is received
and played into the TV set. The set-top box receives the video, decompresses it,
decodes it, and transforms it back into an analog format. Some versions of this
approach call for continued use of the telephone for calling by further
subdividing the spectrum so that incoming and outgoing (full duplex)

conversations can be transferred as well as the upstream and downstream signaling.

Figure 12-8 shows the overall architecture already discussed. The problems with this architecture are not obvious by looking at the concept shown. First, unlike the normal CATV system, where all channels are available at the customer premises, this method of video delivery provides only one video stream at a time, due to the narrow bandwidth available. Thus, if the requester wants to select another channel, he/she must make that request, after which the request message will travel back to the headend and the requested channel will be selected and transmitted to the requester. For the CATV network, the selection is performed at the receiver (TV), whereas for the copper delivery system, the selection is accomplished at the headend.

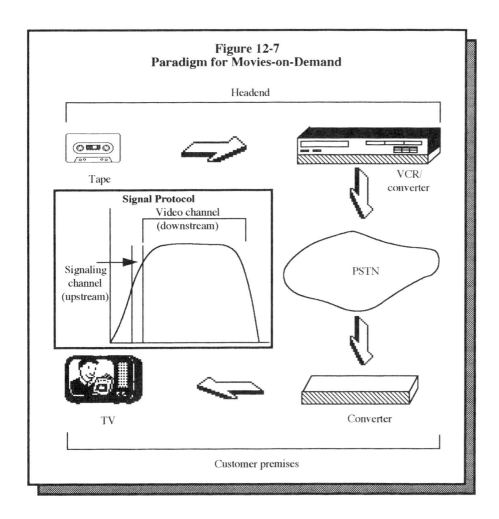

Figure 12-7
Paradigm for Movies-on-Demand

12.4.2   Business and Revenue Opportunities

12.4.2.1   Analog Copper Delivery System

If the architectural approach calls for analog delivery, the immediate conclusion is that the implementation costs will be cheaper. The transmitter and receiver equipment currently cost about 43 percent of corresponding digital solutions. Obviously, this disparity will change in time as digital capabilities began to be produced in quantity, but at present, analog video transmission drivers cost less than $200 each in quantity, whereas the equivalent digital version costs about $700 each. The transmission system requires that the telephone lines be unloaded so that the total bandwidth available to the line be available to carry the signal. This will cost an additional $75 per line to the customer premises.

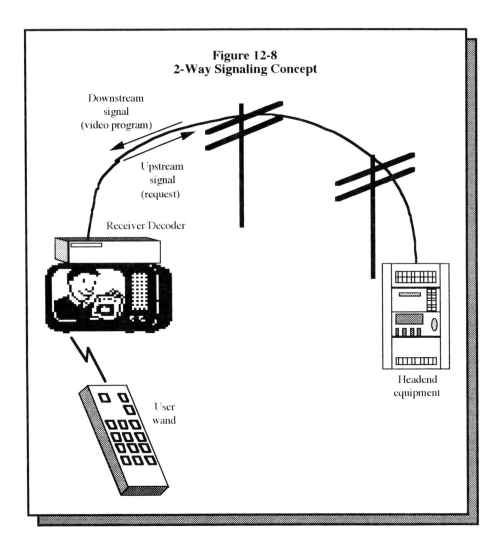

Figure 12-8
2-Way Signaling Concept

The set-top box, needed to handle the requests and condition the signal, costs about $100, also in quantity.

The total implementation cost per subscriber is therefore about $375 for an analog transmission delivery capability. For a per-movie-cost of about $8, of which 50 percent goes to the copyright holder, the remaining margin is $4.00. Over 94 pay-per-view movies must be ordered in order to breakeven, assuming no time value of money, i.e., interest expenses.

12.4.2.2 Digital Copper Delivery System

A digital transmit-and-receive system requires much more sophisticated hardware. Such a 2-way system costs about $700 per transmit/receive pair. The unloading of the telephone line costs about the same as that for the analog system, $75. The set-top unit also costs about $100, the same as before. For the same royalty scenario as described above, the recovery of the implementation cost will require more than 218 pay-per-view transactions. The added cost of the digital hardware includes the cost of the R&D of the compression algorithms, the coding electronics, and the data formatting and transmission components. The transmission lines are likely to require more maintenance in terms of noise control.

12.4.2.3 Analog or Digital Coax Delivery System

A coax network can deliver pay-per-view offerings by either of 2 primary means. One method is via telephone requests either to a service agent or to a computer that can decode touch-tone signals and honor such requests. The other method is for the various requests to be honored, by whatever request method, by sending a signal back down the coax to the requesting set-top box, which will then decode the incoming data stream on a particular channel and allow it to be viewed on the channel specified. This type of system is already in place. The only additional feature that needs to be added are 2-way repeaters so that set-top box requests can be made from the customer premises to the headend control computer.

12.4.3 Network Management Issues

The management of such a system is similar in concept to any other data communications system. The physical management of the system consists of monitoring the interfaces between the provider facility, known as the headend, and the customer premises. The architecture may be either circuit switched or a tree-branched distribution system. The typical CATV network is also a branched configuration that consists of repeaters located periodically along the various branches of the network. Thus, fault detection within this type of configuration will, at best, yield an ambiguity consisting of several repeaters at the very least.   For an analog delivery protocol, there is no link layer involvement, and for both analog and digital delivery capabilities, the network configuration is a combination of both layers 3 and 4.

This type of network also has a requirement for the management and monitoring of the session layer, as well as the presentation and applications layers. The extent to which any of these can be monitored and controlled from a cost-effective point of view is debatable, depending upon the nature of the system involved. The most reliable *a priori* point of management is still the physical layer. Even with a certain amount of ambiguity built into the process of management, the best bet for cost and quality control relies upon the physical monitoring approach. The use of data fusion techniques may also assist in narrowing the possible points of faulty performance, as information from other parts of the network can be analyzed to help eliminate certain potential sources of error.

12.5 Personal Video

Personal video is an offshoot of the video-on-demand discussed above, except that the producer of the material is the private user, not involved in any formal commercial video production and distribution system. The types of scenarios that would support the creation of these types of video clips are family-generated video that portrays family activities and that would be shared with other family members. The use of creation capabilities and distribution facilities would be for the sharing of personal imagery, development of educational material for children, development of presentation material for business and school, as well as other personal needs that may arise.

12.5.1   Application Overview

12.5.1.1   Background

Personal video can best be described as the production of video material for personal use and/or entertainment. It is a market area that parallels camcorders and VCRs in user interest, because without either one of these hardware technologies, personal video production could not exist, and the accompanying transport service could not be argued. In 1994, 17.5 million households in the U.S. had camcorders, and over 75 million households had VCRs; by the year 2000, these numbers will double. These numbers also define the current and projected market for the production side of personal video. The transport side is the transmission of personally produced video to others, who are not in the immediate vicinity.   Such people might be family or friends.   This same technology also has business use potential in such areas as education, insurance, and real estate on the using side of the picture.

The operational scenario would include the attachment of the produced video tape to a network interface, initiation of the connection to the distant recipient, and subsequent transmission of the produced material. Further details of this service will be discussed below.

12.5.1.2   Customer Need or Problem Addressed

Expressing and sizing the need for personally motivated products is similar to sizing the need for the telephone at the turn of the century. Nobody knew exactly what the market need was, and what that market was willing to pay for the services provided. It turned out that there was a big market for personal use telephony, but this was hard to see at the time.

With regard to personal video, however, some correlation can be inferred from the market size for video production equipment and from what is known about media markets and their behavior today. The obvious indication thus far is that there is definitely a market for personal video production, as evidenced by the growth of home movies over the past 30 years. The customer need encompasses not only the production of personal video but also the sharing of that production with others. The production side is expressed by use of camcorders and VCRs to replay personal productions, while the sharing aspect is usually expressed through the use of an automobile to transport the video. An alternative is the use of the mail system/UPS, etc., to distribute the video in question, because the distribution must occur over distances too great for personal transportation to and from the point of display. The distribution of these personal productions can occur by one of three methods, either vehicle, mail, or electronic transmission. There is a time and cost factor associated with each that is weighed by the producer of such material, based upon the urgency and necessity of the sharing contemplated.

There is also a real-time versus a non-real-time aspect of the transmission issue. For non-real-time transmissions, the data rate and compression of such data is not as demanding as for the case where the transmission is at 30 frames per second even at frame sizes of 240 by 320 pixels. The provisioning of a system that is essentially a data scheme would make the delivery of this type of capability much more attractive and easily implemented.

12.5.1.3   Key Features and Capabilities

It is clear that the point of distribution for personal video will most often be the home of the individual producer. Thus, the home network interface should be that starting point. There is some slight possibility that the distribution point could be a series of nodes or access points, such as ATM-like booths distributed throughout the user community, but this adds a level of inconvenience that would probably invalidate the need.

If the distribution point is the home, the user would exercise a set of features that would allow user hookup to the network, probably dial up access, simultaneous voice conversation with the receiving end, and display of the video at the receiving end using some type of interface to the TV set. The connection and its activation for use will have to be extremely simple, as novices are expected to be the recipients for most of such material.

Thus, the capability will be perceived as a 2-way voice system multiplexed with a simultaneous 1-way video signal. This service is distinguished from video telephony in that it uses a 1-way video path. The implementation features will provide for dial up access, simple connection and initiation of the signal to camcorder and/or VCR, coupled with simple display of the signal on a TV, possibly requiring only turning on the TV set that is already interfaced to the phone jack in the house.

12.5.1.4   Benefits

As indicated above, personal benefits are difficult to quantify in terms of monetary value, because revenue displacement and revenue creation are difficult to quantify. However, assuming that the sharing of personal video is a marketable service, either for personal or small business use, such as real estate, the customer will evaluate and utilize the method of distribution that most nearly matches his/her time and expense thresholds. Thus, the benefit is directly proportional to the cost and convenience issues involved. Assuming that the mail system would cost about $2.00 per VCR tape plus cost of the tape which is at least $4.00, the mail distribution method would cost about $6.00, not including transportation, if any, to and from the post office. The personal transportation to and from the point of viewing would be a variable cost that would depend upon distance. The cost of electronic distribution would have to be framed around mailing plus the relatively intangible issue of timeliness. Thus, the cost per distribution usage would have to be on the order of $6.00, plus convenience costs, say $1.00, plus the cost of any long-haul charges, for a total that approximates the mailing cost.

12.5.2   Market View

12.5.2.1   Description of the Target Market and Potential Customers

The total market would include all people who own camcorders and VCRs. Of this total, the addressable portion of this market would include a subset of that whole segmented by discriminates, such as family size, and income level. The target market, which is a further subset of the total market, would include early adopters (usually 3 to 5 percent of the market), and a modest projected growth of that group up through the first 5 years of service offering.

The profile of the target market would most likely include young families where both parents work, disposable income reserves, at least one child, living physically removed from grandparents, and the propensity for technology "gadgets," such as camcorders and information exchange devices.

12.5.2.2   Current and Projected Target Market Window and Size

The information concerning current and projected market windows and sizes can best be expressed as a table that is provided below. These targets are shown as a function of the camcorder market and the market window of 1996 is the service target. The projected market of 32.5M is based on the number of camcorders expected to be in existence by 1996.

	Market Window	Market Size
Current	1995	17.4M
Projected	1998	39.5M

12.5.2.3   Current and Potential Alternative Products, Services, Systems, or Technologies

This section can also be expressed as a table of the alternatives that may be developed and implemented to provide competitive alternatives.

	Alternative Products	Alternative Services
Current	None	Mail Physical transport
Projected	VSAT	Satellite

12.5.3   Conceptual Approach

12.5.3.1   Provisioning Approaches

There are several possible technical approaches to the provisioning of personal video. One is the use of ISDN channels to implement the service identified above. Another is the definition of a specially multiplexed bandwidth carrying both the video and voice conversation that accompanies the personal video component of the session. And a third option is the transport of the 2 different signals over separate circuits, one being used for voice and the other used for video.

The video signal would be carried at baseband frequency, and since the video would, of necessity, be compressed, a better than 4:1 compression ratio would have to be implemented in order to accommodate the bandwidth requirements, as illustrated below. The low-frequency components of the baseband video signal could be attenuated to accommodate the normal conversational voice component.

It appears feasible at this point to assume that a switched offering would be possible, practical, and desirable from an economic point of view. In the case of separate transport for the voice and video, no multiplexing capability would be required. For the case where the 2 signals are multiplexed, muxing (see Figure 12-9) would be required and appropriate equipment would be necessary.

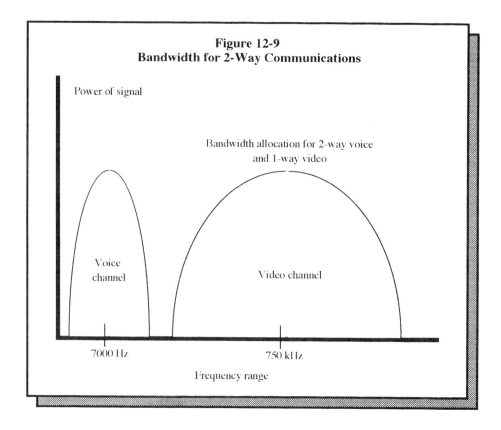

Figure 12-9
Bandwidth for 2-Way Communications

Power of signal

Bandwidth allocation for 2-way voice
and 1-way video

Voice
channel

Video channel

7000 Hz

750 kHz

Frequency range

The operational approach would be to offer the video network service as an enhanced feature of the network, coupling this with requirements for the customer to have a customer premises equipment (CPE) setup incorporating camcorder interface connections directly to the service or through a VCR interface and then to the service.

The composite, multiplexed, digital video signal is estimated to be of T-1 bandwidth or some large fractional part thereof, using a new telco service known as fractional T-1. As a result, the transport of such signals will require a certain technical level of service delivery available to the origination, intralata, interlata, and termination carriers involved. However, the composition and decomposition of those signals, referred to as *muxing*, can be conceived as a function of either the CPE or as a network service. The actual interface of video sensors and conversion equipment is most likely a CPE issue. The use of a T-1 or fractional T-1 capacity to support the personal video initiative is similar to that addressed with ADSL or HDSL. The total bandwidth is divided so as to accommodate signaling as required, the composite video stream, and possible bi-directional video as well. The basic allocation for upstream and downstream signaling bandwidth is illustrated in Figure 12-9. Here, the total spectrum is shown as divided so as to allow video and audio to be conveyed in one direction using 2 different parts of the allocated bandwidth. Extensions to this approach could allocate an additional portion of the bandwidth for signaling.

12.5.4   Key Issues

12.5.4.1   Possible Regulatory/Legal Constraints

As an enhanced service, personal video will have to be appropriately isolated from basic services and pay for the usage of certain basic services that support the provisioning and delivery of the personal video capability. The usual complaints from the CATV industry will undoubtedly be raised, but such complaints would have to be based upon issues such as whether the CATV industry has similar capabilities to provide such a service, and whether this service infringes upon existing revenue streams of the CATV industry. The answer in both cases appears to be "no" at this point.

12.5.4.2   Technical Efforts Required

The following technical efforts must be accomplished prior to deployment of the service:

(1)   Demonstration of the concept

(2)   Initial and growth market sizing

(3)   Development of the operational concept, functional requirements, and top level, as well as detailed, specifications of the system and its components

(4)   A field trial to reflect both the market and technical feasibility

(5)   An initial and limited offering in, hopefully, each region

12.5.4.3   Special/Unique Client Capabilities Needed

In order to implement this service offering, the regional companies must be able to start and operate a business of this type. This implies that these companies have or can acquire the necessary program development expertise, can exercise the necessary systems engineering to oversee the architectures, can exercise the correct business efforts to partner with the right companies, and can provision the developed system.

12.6 Personal Network Directory

Studies conducted by Bellcore for the regional telephone companies in the U.S. have confirmed that the greatest source of telephone numbers for personal usage by telephone subscribers is their own personal telephone directory. This can range from scraps of paper, to a note pad, to elaborate Day Timer address books, to sophisticated electronic address books where the user may have phone numbers recorded for private use. Discussed in this section is a method that can organize the phone and address lists for all diverse user groups into a format suitable for and important to each of the various interests and needs of the user cross section. This product comes in the form of your own personal phone list periodically updated and organized, and then printed and bound in a form that is easy to carry and use. How this is done and what its implications are will be discussed below.

The concept of the personal directory has already been explored by most of the large telecommunications companies, in some cases as far as public trials. The advent of the palmtop computers and other devices, such as the miniature address and phone number calculators, has made the personal directory, in its full-fledged form, only a short time away.

12.6.1   Application Overview

This section will provide a conceptual view of the structure and operation of the personal directory business development opportunities. Referring to the personal directory as a business also includes its engineering and developmental content. The personal directory as a tool is, in effect, a channel for advertising dollars which are exchanged for a very targeted and segmented population of potential customers. The customer perceives the directory to have value based upon its utility in his/her lifestyle. The directory consists of two major pieces of information, namely, directory listings of selective interest to the individual, and advertising targeted for that customer and presumably of interest to that customer.

The product (directory) is generated by permission from the user to the telco to use his/her CPNI (customer proprietary network information), i.e., his/her called phone numbers, in order to essentially "reformat" his/her monthly phone bill information, and to give it back to the individual with advertising in a convenient pocket or purse sized, booklike form. The user, therefore, becomes a subscriber to a free service paid for by advertising dollars, and in turn receives something of value.

The personal directory product opportunity is paid for in 3 ways:

(1)   Revenue from advertising

(2)   Increased utilization of the network from incremental phone usage

(3)   Eventual incremental revenues from calling card usage, transactional fees, directory customization charges, etc.

Thus, the personal directory, like the Yellow Pages books, is paid for by advertising, but a more targeted type of advertising where lifestyle is an indicator of products and services likely to be of interest to the affected subscriber.

While it is true that there appears to be a potential for cross-elastic competition effects, that is, shared usage between the 2 types of products, the personal directory and the traditional Yellow Pages book, the personal directory in fact targets narrow groups of users, not the consuming public in general.

The personal directory, in its initial form, will not have all of the multimedia features that the final version will contain. Later in this chapter, Figure 12-13 will illustrate how this concept becomes a full-fledged multimedia capability for both the user and provider.

12.6.2   The Market

12.6.2.1   Historical Perspective

Personal mobility and an individual's diversity of needs is a fact of life as we approach the year 2000. The explosion of information, in terms of type, quantity, and availability, has created a complex and sometimes confusing set of problems in terms of how we deal with and use information. The multitude of our choices is many times overwhelming, as is our exposure to facts. Nowhere is the confusion of choices and need for selectivity more obvious than in choosing nearby home-convenient services and/or stores.

The attempts to solve such problems have resulted in the investigation of many new technologies to assist in developing the time and choice management products and advertising services which have invaded the commercial marketplace over the past several years. Such products have attempted to take advantage of leading-edge technologies, such as circuit microminiaturization, computer-assisted design, application-specific integrated circuitry, and the various kinds of communications (digital and tone) technologies.

Not surprisingly, but in a parallel and somewhat unrelated fashion, the printing and publishing industry has witnessed an infusion of new technologies. These technologies range from bar-coded information to revolutions in graphics design, layout, and composition, from computerized typesetting to the database integration of text and graphics. Personal management products and advertising services incorporating such technologies range from fill-in address books and calendars to bar-coded information, holograms, and sophisticated, computer-assisted time management tools which interface to personal printers. All of these advances in diverse technology arenas are about to culminate in an explosion of personal lifestyle management.

The idea of time and information management is not new, however. For example, in 1855 the town of Newburyport, Massachusetts, published a directory listing of the town that included names and addresses of the residents (without telephone numbers, of course), and provided in the book such extra tidbits of information as almanac data, a list of local societies in the area, an area for the subscriber to record notes, and, not surprisingly, a section for advertising by the merchants in the town. This book was published in the days before the telephone, but was obviously the forerunner of the current white and Yellow Pages directory. It is clear by browsing through this document and any other like it that time and information management was an issue of concern to people even 100 years ago.

12.6.2.2   Recent Events

In recent years, there have been a wide variety of "tools" made available in the marketplace to help us manage the choices available to and pressing upon us, as well as the time required to do justice to such choices. Examples of such handy aids, in electronic form, are calculatorlike devices which store and retrieve names, addresses, and phone numbers. These products go by several copyrighted names such as "Calling Card," "Travelmate," "Little Black Book," "Digital Diary," and "Business Card." The paper form of similar products are "Daytimer," "Filofax," and "Executive Scancard." Hybrid forms, meaning combination electronic and print versions, of personal management systems go by names such as "Time/Design," and "Dictaphone." In all such devices, however, the individual has to "input" or transcribe the information into the medium on his or her own.

12.6.2.3   Problem/Opportunity

What's the problem then, and what's the opportunity for the customer base? There are two problems involved here which lend themselves to a synergistic solution. First, people are very reluctant to pay for information, even if it is of interest to them. Second, if it is provided free of charge, it soon becomes something which is craved and becomes habit forming. The evening news programs are good examples of this assertion.

There is currently no suitable parallel to the business being proposed here. The closest things to it are the various paper and electronic directories which currently inhabit the marketplace. Such products require that the user purchase the product and then expend time filling in the information of interest.

What is being proposed is a product which is provided free of charge, complete in content, and tailored for that individual's potential purchase needs, e.g., nearest pizza parlor, local promotions, etc. It is a multimedia, interactive system in that each individual interfaces with a central updating repository to specify his/her choices. Such buying decisions many times incorporate motivations, such as snob appeal and/or status recognition. The trend in advertising today is toward shorter-term ad campaigns to capitalize upon product diversification and/or product improvement discoveries, better targeting of the customer base, and shorter advertising response times due to budgetary constraints.

The opportunity is essentially an amalgam of the direct marketing and personal information management businesses, each of which is at the mature end of its respective development cycle, and is coupled with the inclusion of value-added features provided by the telcos or third-party providers that offer the service. The following are the essential advantages for all concerned.

(1)   The consumer gets his/her personal directory on a periodic basis, automatically updated for him or her

(2)   The provider gets revenue from advertising and certain other incremental revenue increases and the telephone system may benefit to the extent that the subscriber uses the network more

(3)   The advertiser gets a more segmented population from which to target advertising dollars

With this approach, the consumer has access to the products and personal service needed while having the cost of the solution paid for through advertising. In this way, the advertisers pay for the privilege of getting their message in front of the consumer, thereby paying for the expense of providing a valuable service to the customer, i.e., a mechanism for the customer to manage his/her information more productively.

The opportunity, therefore, is to provide something of value to the consumer, to which the advertiser will want access and which will pay for the costs of collection, publication, and distribution. As mentioned above, the Yellow Pages have for some time begun to be fractured into different offerings, as there are boaters' Yellow Pages, ethnic-language Yellow Pages, neighborhood Yellow Pages, and the like. Of course, the ultimate target of such vertical and stratified attacks on directory markets is a book suited for the specific individual. The personal directory, customized for the individual, and providing the features which make the directory of value to the user, incorporating advertising to offset costs of production and distribution, is the ultimate objective market segmentation, and therefore an opportunity for the telcos, third-party directory companies, or even other third-party providers of printed material.

12.6.2.4   Product Potential

As costs of providing products and services to the public continue to increase, the opportunity to provide a product which is partially or totally paid for through collateral revenues becomes more and more attractive. The collateral revenues to be generated by this product offering are from advertising, processing of transactions relating to purchases, and premium-level information extractions, storage, and retrieval. The personal network directory could, in its ultimate form, expect to displace direct mail revenues and, in phased releases, to generate revenues from the exercise of transactions, both voice and data, over the public network.

While it is beyond the scope of this treatise to derive market projections for the personal directory, the market potential for this product is expected to be very large. The general logic for the market potential, however, goes something like this. Typically, the percentage of early adopters for high-tech products and services can be expected to be as high as 5 percent. If only 1 percent of the total 200 million or so telephone subscribers were to subscribe to the personal directory, its market would be 2 million. And if the revenue for each subscription were worth at least $10 per month to the product provider, the dollar value for this directory would be at least $240 million. Other factors that must be taken into consideration, of course, are the timing of the offering, the introduction of the offering, the ramp-up of the product subscriptions, the degree to which off-the-shelf components can be utilized, and the delivery costs that would be incurred.

The market potential for the directory must be predicated upon the trends in the advertising world, the fractured products offered in the marketplace, and the mobility of the consumer population. All of this adds up to a need which can be addressed by the personal directory.

The market potential described here is distinctly separable from the Yellow Pages currently published by the various regional and independent telephone companies. In order to prove this point, we shall draw upon the scientific method. According to this method, we must initially assume that there is absolutely no difference between the Yellow Pages and the personal directory, and then show that, in fact, there are differences. We can only show this by pointing out enough differences to satisfy the observer that, in fact, there are numerous dissimilarities between these businesses. For companies already in the Yellow Pages business, the possibility of revenue dilution from their core business of traditional books may prevent them from launching into this new venture for fear of diluting their current profits. For companies that are not already in the business, this reluctance by the current group of providers could be a great incentive and level of protection while they are becoming established.

This distinction can be made by noting the following:

(1) The Yellow Pages are annual; the personal directory, on the other hand, is something short of that, possibly monthly in nature.

(2) The Yellow Pages carry ads which are long-standing and have tendencies not to fracture into new products during the cycle period; the personal directory carries short-term ads about products which tend to migrate or evolve quickly.

(3) The Yellow Pages are global in their coverage of a specific region; the personal directory covers the needs of a specific individual.

(4) The Yellow Pages have no coverage of personal time and information management; the personal directory is about the "cloud of numbers" which follow each individual and how those numbers are managed.

12.6.3   The Product

12.6.3.1   Consumer End Product

The ultimate product which is to be put in the hands of the individual is partly a paper document which, in the first phase, contains only 2 sections, namely, personal phone numbers which have been reversed out from the telephone call accounting database to yield names, addresses, and phone numbers of the called locations, and advertising material which is targeted to that individual and therefore perceived to be of value to that person.

The second major part of the personal directory is its keypad/display module. This second feature is important in allowing the user to capture other numbers, to access remotely database items of interest to the user, and on-line display of numbers, plus call setup using the RJ-11 phone jack built into the system. The unit would probably have a bar-code reader so as to facilitate the capture of numbers and other information. These locally captured bits and pieces of information could periodically be uploaded to a central repository for consolidation with other data and subsequent relay back to the individual's unit for hard-copy and/or electronic usage. Image scanning could also be added to the electronic version for the capture and integration of imagery into the system.

In later phases, the document/electronic keypad system might contain additional sections of information, such as those for voice gateways, data services, and/or videotext capabilities. On the inside front cover and at strategic places within the document, color ads might be placed. In even later phases, depending upon the acceptance of the medium, the personal directory might even be positioned as a "personal magazine" in which short articles or informational tidbits might be inserted. Since the document would be subscriber based, it could also be a source of emergency information for the individual, available as necessary for health care reasons. On the first page, for instance, the individual's personal information could be provided which might consist of several types, such as numbers of interest, names and phone numbers of agents, e.g., lawyer, insurance, accountant, etc., and/or family information. Succeeding pages would be interspersed with ads, items of interest, such as current events calendars, seating charts for public facilities, maps showing references to services of interest to that individual, general information, informational sources, personal calendars, notes, etc.

The major value to be added to the personal directory is a sorting and reformatting of the user's most-called phone numbers over some period of time, which are printed and organized in the directory. This feature will most likely require permission from the individual concerned for phone company release but will be part of the sign-up process for the directory itself.

12.6.3.2   System Architecture

The personal directory is the perceptible part of a set of products and services which comprise the total of the personal network directory system. The system has several functional elements which comprise its totality, among which are a service center, a publishing facility, information storage and retrieval capabilities, an advertising interface, and a transaction functionality.

12.6.3.3   User Interface

The physical object used by the consumer is a book, which is part sectionalized print material and part a simplified electronic calculator lookalike. The basic scenario is that the user will carry the directory in his/her shirt pocket or purse or other convenient place on the person. The directory will be used on occasions where references to telephone numbers, appointments, notes, or other information need to be made.

The sectionalized print material will consist of personal information, notes, business cards, telephone lists, appointments, and advertising material. The final form of the directory, in which a touchpad capability is present, will facilitate dialing by generating calls from stored phone numbers that comprise part of the user's personal directory database.

The electronic feature of the directory will consist of a keyboard, memory, display, and external interface(s). These interfaces may include an RJ-11 phone jack, a bar-code reader, and possibly a scanner. It will have the ability to receive, store, and recall information entered directly into it via the keyboard. It will also have the ability to send and receive information via its external interface facility. And it will have the ability to provide a conversion between electronic form and hard-copy form. All of this is in support of updating the print sections of the system, or to supplement it.

12.6.3.4   Operational Scenario

The system has several modes of operation, which include telephone list usage and maintenance, information storage and retrieval in the form of calendar updates and notes, and order entry in the form of purchases and data query.

The typical operational scenario would involve a person making a call from his/her office, being plugged into the public network via the directory, and using the directory to initiate and log the call. This same connection would also be used to pass a message to an associate at another location, and to send a business card to someone across the region.

Later on, the person would make a call from the airport, where he/she was about to catch a flight. This activity is effected by using the tone-generation capability of the directory to dial the number and record the call. The user updates his/her calendar based upon the call, using the keypad entry capability of the electronics. These data are transmitted to the service center for later integration with a new calendar update by the user's initiation of another call and sending the calendar update digitally.

Finally, after getting home or to his/her overnight rest stop, the user plugs into the phone and initiates a call to the service center to make an order for goods or services. This activity is accomplished by electronically dialing the service center number, and completing the transaction using the keypad to effect this activity.

12.6.4   The Technology

There are several implementing technologies which support the personal directory functionality and its system design. These technologies have been arrived at after performing a detailed functional analysis of the system requirements. Both the system design and the personal directory functions are also products of the usage requirements.

12.6.4.1   System Design

The personal directory system has three major physical segments associated with it. These are the service center, the distributed publishing facility, and the personal directory itself, which comes in many copies.

The element which supports the personal directory in terms of maintaining personalized information, completing transactions, and integration of advertising material is the service center. This is analogous to a data processing center which is on-line and in communication with users of the personal directories through the telephone system. The users of such a facility will be using it to query the service center for information, such as stock reports, or to provide personal information which is to be issued in the individual's next directory update, or possibly to conduct a transaction related to shopping or information purchases.

The service center is a transaction processing and database management facility which has information related to the topics of center operations, such as lists of suppliers and information on their products, detailed descriptions of vendor products (possibly in catalog form), supplier marketing requirements relative to the personal directory business activities, such as target numbers, costs, and results, customer profile information, sales histories, security control, and transaction control data, such as credit card checks, etc. The service center also has other functions to perform, such as the on-demand generation and transmission of business cards, postcards, greeting cards, etc. All such features are provided as a result of transmissions to/from the directory user and the service center.

The publishing facility, known as the distributed publishing operation, has a remote communications interface with the service center and the necessary monitoring and control capabilities needed to temporarily hold and queue up the print requirements for each directory, control of the actual printing activity and the binding/bundling as well. In some sense the publishing facility can be construed as a set of dispersed printers which have a close proximity to the consumer.

12.6.4.2   Key Technologies

The concept of a product that can be produced on a potentially massive scale and distributed to 2 million or more people and be individualized for each one of these people presents several problems which must be accounted for in the system design. Such a problem is not, however, entirely unfamiliar to the telephone companies, which, for the larger telcos, require that they produce between 10 and 20 million or more telephone bills each month, all of which are personalized.

There are several technologies which are key to the success of this undertaking. These technologies include the following.

(1)   <u>High-Speed Printing:</u>
Reasonably high-quality printing, of 240 DPI or better, at high
speeds

Description -
More diverse applications will be imposed upon the printing function as
distributed publishing and printing become the accepted mode of
operation. These applications require rapid response and variability in
the capability as such technology applications as "books on demand,"
hypertext, etc., come of age. The personal directory is a step into this
realm of personalized media products.

Requirements -
Requirements of at least 5 million feet per month will soon be a reality of
either sheet or roll feed with resolutions of at least 240 by 240 DPI or
better. 300 DPI is a more acceptable standard because of compatibility
with scanning technology resolutions. Duplex printing (both sides of
the paper) is a definite requirement also. The printing device must be
capable of handling mixed media, i.e., text, graphics, and imagery as well,
hence it must be bit mapped. The required standard of implementation is
the Adobe PostScript standard. Finally, the print system must be
capable of both gray-scale as well as 4-color reproduction.

(2)   <u>Color Printing:</u>
Color reproduction of good quality which is available at low cost
and which can be easily integrated into the paper stream through
specialized collating machinery during the binding process, and
capable of handling at least 5000 ad reproductions per run

Description -
Publishing competition will sharpen around such innovations as color
renderings and enhancements of illustrations. The technology is
improving to the extent that high-quality color imagery can be expected
at low cost in the near future.

Requirements -
Four-color scanning and true reproduction will soon be absolute
requirements. Pseudo-color derived from gray-scale imagery will continue
to be an important adjunct technology, however, for color renderings.
Several techniques are being investigated and productized for the actual
color imprinting. Among these techniques are liquid crystal light
illumination to produce 128 gray-scale levels which in turn are exposed to
light which casts a variable projection upon the paper to expose it. The
variability of exposure determines the color scale position. The trick
to such technologies is the resolution as well as the veracity of the color

tinting. To date, color resolutions of 300 DPI or better have been achieved and should be a minimum requirement for future capabilities.

(3)   **Cost-Effective Binding System:**
A binding process which is low cost, capable of handling 9 duplex personal directory pages per 8 $\frac{1}{2}$ by 11 sheet, and easily integrated into each periodic update of the directory

Description -
The mechanical folding, cutting, collating, and binding of books of varying sizes is a key issue for future considerations in the publishing business. Specialty publications, such as "books on demand," similar to the personal directory, will require high-speed and high-quality capability. The personal directory will require special sizing in order to meet the needs of different individuals.

Requirements -
The binding systems required for future applications must accommodate at least 36,000 sheets per hour, which in turn will be folded, cut, and collated into 36-page documents. Current binding systems can accommodate 5000 books per hour. These cut stacks will then be passed through stations which add front and back covers, are stapled and bound and banded. The banding will be handled by mailing labels which correspond to booklet groups and are attached to books which are printed on a predetermined basis according to locality.

(4)   **Distributed, Intelligent, Multimedia, Object-Oriented DBMS:**
A database system which has several capabilities, such as mixed media storage of graphics and text, an inventory tracking capacity, interactivity with the administrative dataset (CADIS)

Description -
The database system is the cornerstone of most computer systems. The system necessary for publishing applications requires mixed-media storage capability so that text and graphics information can reside in the same environment. Additionally, the data should be in the same or compatible format so that extra conversion steps can be avoided in the output stages.

Requirements -
The multimedia database requires that images be stored in formats which accommodate at least 256 color tints, at 300 DPI, and span at least an area of 8 $\frac{1}{2}$ by 11 inches. In addition, the database accesses must be distributed, and the database must be object-oriented, such that subscriber-oriented information can be easily associated and "stacked" for ease of

printing. The database system must be capable of storing and outputting formatted mixed text and graphic information for printing.

(5) Scanning/OCR:
High-quality image scanning with OCR capability and format compatibility with the database system

Description -
The scanning and interpretation of images and text is a potential area of exploitation for the personal directory. Scanning can be a source for personal image collection, and interpretation can be a means for aiding in making judgments about the imagery. Text, in particular, can currently be interpreted from many font styles and pitch sizes. This capability can increase productivity by saving time that would ordinarily be consumed in retypesetting text information. Scanning is especially important in the high-quality capture of imagery.

Requirements -
The scanning devices of the future must be capable of capturing imagery of at least 600 DPI or better. The OCR functionality must also be capable of text capture accuracy of at least an 85 percent success rate for the 50 most popular font styles ranging in pitches of 9 point to 30 point, proportionately spaced, and in various styles, e.g., bold, italic, underlined, etc.

(6) Image Processing:
Image cropping, retouching, smoothing, and rescaling capabilities

Description -
This capability is critical to certain applications requiring imagery as opposed to graphics. The ability to invert, rotate, blur, sharpen, retouch, repaint, etc., help to alleviate the necessity for reshooting certain scenes. The personal directory will be loaded with imagery, and therefore this technology is key to its success.

Requirements -
A functional workstation which has the capability to manage the control, manipulation, and alteration of images is required for systems of the present and future. 2048 by 2048 displays will be required which can display an image and scroll and/or pan across its entire area. This capability is required because even such high-pixel-density displays cannot accommodate the minimum display resolution of 300 DPI without the pan-and-scroll feature. Such a system must also have the feature of showing the imagery at several

zoom factors, such as normal size, with increasing compression factors
down to a level of at least 8 to 1.

(7) Computer-Assisted Layout and Composition:
Computer-assisted information integration, database extraction,
page placement, and pagination

Description -
The anticipation of "books on demand," such as the personal directory,
makes it necessary to plan for the automation of certain features related
to the extraction of information which will be organized for composition
within the guidelines determined by the layout software. Such a
functionality requires a rule-based system which can assist in the
decisions of page sizing, illustration placement, proportionality, and the
like.

Requirements -
This functionality must be capable of at least 250 rules, such that the
intelligence can be adequately embedded into the functionality. The
requirements are generally to size the space needs of the various sections
of the document, and based upon such, to make decisions about
priorities and to rescale print requirements as necessary to accommodate
overages of data output.

(8) Personality Profile Generation:
Personalization of specialty information enclosures and advertising
profile targeting

Description -
The ultimate value of many applications of advertising material is
whether the ads are properly directed at the right audience and, in
particular, the right person. The ultimate aim is to target the individual
based upon not only his or her statements, but also upon actual practice
as well. Demographic data, coupled with customer expressions of life-style,
and habits should be factored together to yield a better picture of the
consumer's likelihood of interacting with this or that topic. Such topics
are many times related to products or services or to information.

Requirements -
The requirement is to provide a capability which matches personal data
with buying habits and/or preferences, and demographics to yield a set
of profiles. These profiles would then be assigned varying degrees of
propensity toward categories of need.

(9)   Data Compression:
Compressed storage and transmission of mixed-media, multicategorized personal directory material

Description -
The position of data as a revenue support and revenue generation tool underscores the entire set of issues related to data compression. Not only is data compression needed to support storage problems, but it is also needed to support wideband width transfer requirements.

Requirements -
Data compression technology becomes more and more important as electronic transfer of imagery is used to support such applications as distributed publishing and electronic publishing needs. Compression ratios from 8:1 to 200:1 are possible depending upon the nature of the information and its format, e.g., raster, vector, etc. Minimum requirements of at least 100:1 should be set in order to deal with data transfer needs in the foreseeable future, at least until broadband transmission becomes a reality on a regular basis. Even then, the demand will probably surpass the capacity of broadband capabilities at that point.

(10)   Transaction Control Support:
Facilities to handle requests across the directory-to-service center interface, as well as messages and responses to messages

Description -
The combination of digital, machine-interpretable queues and commands, coupled with ease of use, e.g., touch-tone telephone, are the setting for transactions from remote locations supported by the telephone. The use of such capabilities facilitates the exercise of information and product/service needs of customers. If such uses can be integrated into print-based systems, the power of the revenue opportunity increases tremendously.

Requirements -
Interactive information gathering and transactional activities should be possible with minimum capabilities of at least 30,000 automated nodes which can recognize and interface with incoming calls for information or services. The system must be able also to schedule as well as identify such needs and to automatically back-order or bring up new resources to support increased requirements.

12.6.5   The Issues

There are two major aspects to the successful introduction of the personal directory into the marketplace, product positioning and the perceived value of the product. Other issues related to this effort include update cycles for the directory, distribution channels, electronic data interchange (EDI), voice mail and e-mail, etc.

This product will be positioned as an aid to consumers which will assist us to track and maintain the array of numbers that follow us around in our daily lives. This positioning puts this system in the information provisioning business.

The fact that we may provide the consumer additional information of interest, such as via the on-line data services, or entertainment and sports contacts, through per-minute charges for audio text-style services, is part of the normal and ongoing service provided by any telecommunications facilities. The difference here is that the personal directory is personalized for each individual user. Thus, the directory is assumed to have value in this regard as well as that derived from frequently called phone numbers being reprinted in the directory for each individual. Future releases will have postcard, business card, voice mail alerts, e-mail messages, and greeting card interchanges.

The update cycles for this directory will vary depending upon the product release phase, but will initially be no less frequent than once per month and eventually anticipated to be once per week. The distribution channel for this directory is assumed to be via several different methods, one being via the mail, either as an insert to a normal monthly bill or by itself, or by both methods. The second major means of distribution will be through electronic transfer from the service center direct to the user's unit over the existing telephone lines. The inclusion of voice mail, e-mail, EDI, etc., will also be accommodated through an electronic interface between the directory and the central service center facility, which will turn such messages around into print and redistribute them to recipients via the next directory update.

12.6.6   Product Definition

The users of the system are referred to as "subscribers," because private customers specifically request this service in response to solicitations made through media channels to the prospective subscribers. The advertisers are referred to as "clients," who provide the revenue to produce and distribute the directory.

The personal network directory will be a convenient, pocket-sized document, half hard-copy and the other half being a PDC (personal directory computer). The hard-copy portion will measure approximately 3 by 5 in., initially containing about 36 pages which will include personal telephone listings, general-interest data, calendar, and advertising packaged inside an attractive, flexible cover and with the individual's name on the front. The PDC will have about the same dimensions as the hard-copy section and will have external interface devices, such as a bar-code reader, image scanner, and built-in modem.

The document will be updated once per month, from information derived directly or indirectly from personal telephone bills of the subscribers, and information indicated at time of subscription, such as expressions of interest in sports, music, etc. Advertising will be generated from clients who are active in the direct marketing business.

12.6.7   Service and Support

The user, referred to as the subscriber, will initiate his/her own subscription by responding to solicitations provided in print advertising or possibly radio and TV ads. The subscription process is intended to be as painless as possible, collecting only the barest of information from the customer, such as name, permission to send this document, interests (sports, music, flying, etc.), and some indication of habits (business travel, hobbies, etc.). It is anticipated that the subscriber responses will be via 800 number call-in to initiate the process or to request the subscription form.

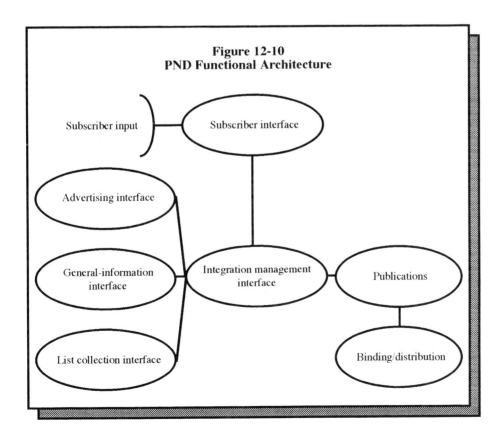

Figure 12-10
PND Functional Architecture

This information will be input to a database which will form the basis for general-interest information inclusions into the directories, and customized advertising. General information will be derived from general sources, such as theaters and sports complexes, while advertising will be sold to key high-end clients, such as the credit card companies, brokerage firms, greeting card companies, and the like. The general information and the graphics pertaining thereto will be input into the database system via desktop graphics capabilities or scanners which support PostScript output to compatible printers.

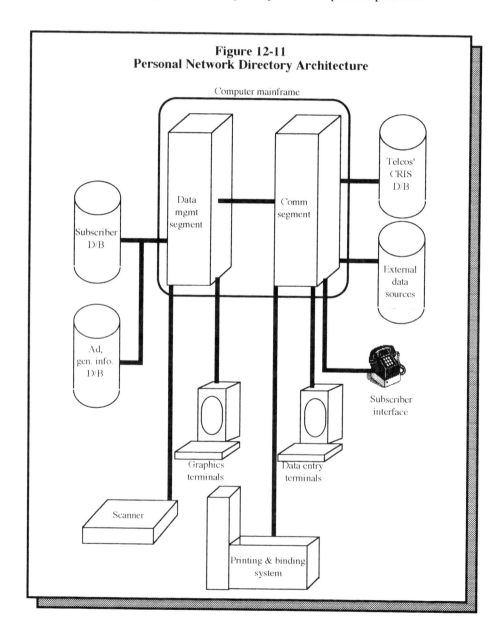

Figure 12-11
Personal Network Directory Architecture

The operational software will collect personal phone information from the telephone call-completion databases, and user-submitted information will form the basis of personalized advertising to be targeted at the user. This assembled information on each person will then be integrated into a file which in turn will form the dataset for use by the publishing software. The publishing software will then compose the subscriber's personal directory book. The composed document will finally be sent to the printing and binding machinery, out of which the distribution will take place.

12.6.8 Architecture

The system functional architecture, presented in Figure 12-10, can be thought of as consisting of the following elements:

(1) The advertising account management element whose tasks are to provide sales support, marketing analysis, revenue tracking, and advertising information capture

(2) The general-information clearing operation element that collects, edits, and organizes general-interest information to be included in the subscriber's editions on a selective basis depending upon whether the user has requested a certain topic for inclusion

(3) The list-collection activity element which searches and sorts through data files for required phone list entries

(4) The integration management facility element which collects all of the information provided above and organizes that into a coherent package for each subscriber

(5) The publishing operations element which perform the layout and composition of the integrated package

(6) The printing operation element which takes the assembled and composed package and then prints, binds, and finishes the packages

(7) The subscription interface element which initiates and establishes a place holder for each subscription which is started or stopped and provides an interface for the input and storage of subscriber data from which subsequent queries are made by the integration management facility

The operation will be broken into 6 major processes:

(1)   The subscriber support process

(2)   The general-information sourcing process

(3)   Advertising process

(4)   Telephone list process

(5)   The data integration process

(6)   The publications/binding process

The subscriber support process will function to provide subscriber information collection and data entry. The data integration process will function to parse information from various databases and to package it into a form for the composition activity. The publications and binding process will take the packaged set of data for each subscriber and print, bind, and finish each personal network directory.

Figure 12-11 shows the system physical architecture within which reside the communications and data management segments, both of which in turn support the 6 processes involved in making the system function. Figure 12-12 shows the operational organization of the system and how the 6 processes interrelate.

12.6.9   The Market

Personal mobility and diversity of needs is a fact of life in the 1990s. The explosion of information, in terms of type, quantity, and availability, has created a complex and sometimes confusing set of problems of how to deal with it and how to use it. The multitude of choices is many times overwhelming. You don't have to go far to see the myriad of selections open to us in entertainment, banking, investment, insurance, travel, etc.

The attempts to solve such problems have resulted in the investigation of many new technologies to assist in developing the time and choice management products which have invaded the commercial marketplace over the past several years. Such products have attempted to take advantage of leading-edge technologies, such as circuit microminiaturization, application-specific integrated circuitry, and various kinds of communications (digital and tone) technologies, as well as calculator-style address books.

Many print technologies have also been introduced into the marketplace. Products incorporating such technologies range from fill-in address books and calendars to bar-coded information, holograms, and very sophisticated, computer-assisted time management tools that interface to printers, to name a few.

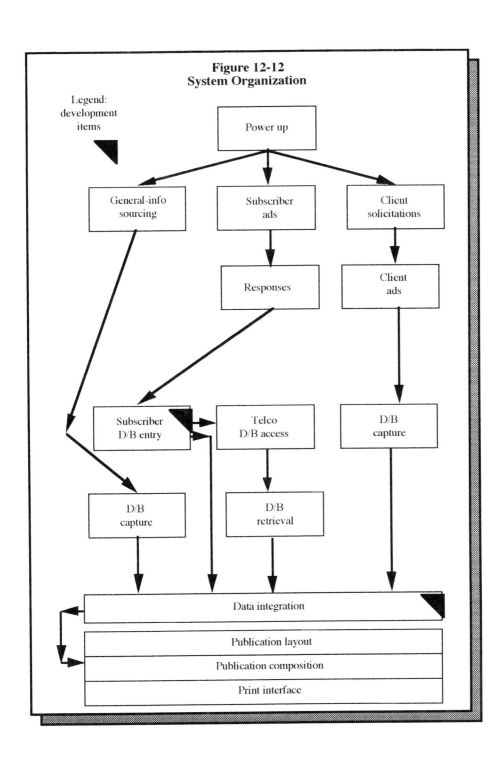

Figure 12-12
System Organization

12.6.10   Risks and Justification

The major risks assumed for such an undertaking as this are described below.

(1) There may be insufficient interest on the part of the public for the directory to attract a sustaining vendor/supplier base. A corollary to this risk is that the directory may be directed toward the wrong market segment. If the sustaining costs for the service center are $50k per month for the processing system and the operations and maintenance (O&M) costs are assumed to be equal to this cost, and assuming that the publication costs are 1/3 of the total system costs, then the publication will require $600k. If each book can be produced for $1, then the total population required to sustain the system is 600,000 customers. If the advertising budget for the aggregate supplier pool is 3 percent and the requirement is to increase margins by 1 percent, for return, then the total sales required as a result of the system is $15 million per year. This means that each consumer carrying around a personal directory will have to, on the average, spend $25 in order to justify the system costs. The spending requirements will go up in an assumed linear fashion, within some limits, as each percent of increased margin is required. Thus, the risk is whether the advertising costs can be justified.

(2) There are several varieties of technical risk which could manifest in providing a defective product. Some of these include inability of the bar-code input and translation from being operable enough to satisfy the need, inability of the system to effectively provide the turnaround time necessary to support calendar requirements and other periodic needs of the customer base, and inability of the system to produce and store the data necessary for each customer at target costs. The issues surrounding costs hinge upon what the consumer will sustain or what the advertisers will sustain, because even though most of the cost of operation will be supported by advertising, some of the cost will also be borne by the consumers as they require varying amounts of individual support to meet their personal directory needs.

The personal network directory as described in this plan is implemented by a listing agreement between the user and the operating entity. The service center operator will in turn collect, store, and redistribute sets or subsets of the information provided by the user for printing, binding, and distribution back to the user.

12.6.11   Description of the Personal Directory

As illustrated in Figure 12-13, the personal network directory consists of 2 major segments, the hard-copy portion and the electronic portion. The electronic device segment contains several interface components, the bar-code reader, the scanner, and a built-in modem. All of these components may be configured so as to be contained entirely within the device framework, thus eliminating the need for extraneous cabling except for the telephone interface. The following is a list of the anticipated electronic technologies that would facilitate the use of the personal directory, electronic form:

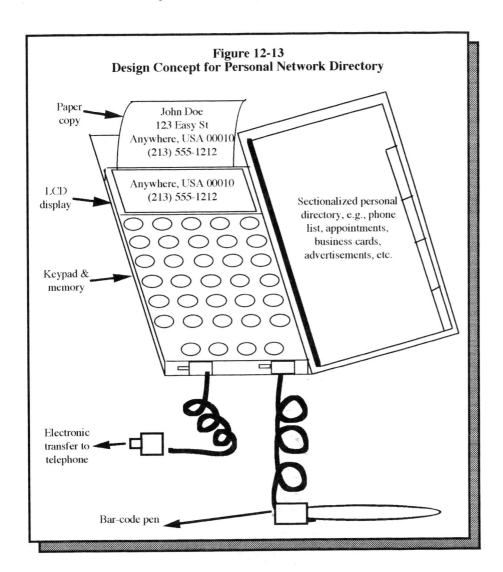

Figure 12-13
Design Concept for Personal Network Directory

Paper copy

John Doe
123 Easy St
Anywhere, USA 00010
(213) 555-1212

LCD display

Anywhere, USA 00010
(213) 555-1212

Sectionalized personal directory, e.g., phone list, appointments, business cards, advertisements, etc.

Keypad & memory

Electronic transfer to telephone

Bar-code pen

(1)   Bar-Code Reader -
Used to read telephone numbers, and addresses for storage and later uploading to the central processing repository.

(2)   Document Scanner -
Used to scan imagery and text of interest to the subscriber. Such stored scans may be uploaded to the processing repository for later printing and return to the subscriber in his/her subsequent directory

hard-copy outputs, or may be retained in memory for later printout on the user's printer.

(3)   **Modem -**
Used for communications interfacing between the personal directory and the central repository facility.

(4)   **Display Screen -**
Used to display pertinent textual directory information and graphical imagery. The display may be a 280 by 320 displayable area.

(5)   **Keypad -**
Used to enter textual information directly into memory. The keypad will contain at least all alpha characters, plus the following additional keys: *, #, /, @, (,), &, %, etc.

The user could receive periodic personal directory updates either in hard-copy form or by electronic update that could then be used either by recall from the display or by printout for reference in hard-copy form. The user could also collect his/her own information, either text or imagery, by bar-code scanning or document scanning for usage directly by the user, or made available for upload to the central processing facility for integration with call accounting information that is collated and relayed back to the subscriber for local usage.

Summary

There is a big future for multimedia systems, especially ones that will utilize the power of networks in order to achieve their objectives. The challenge for providers is to design these systems so as to be entirely seamless and transparent to the users of these systems. Private network developers should also be aware that their implementations can influence the success of their company operations as well. The primary tool for those examples cited in this chapter are those that begin with teleconferencing capabilities, and then leverage other data facilities in order to round out the desired results. The needs of developers include effective media storage systems, high-speed packet transmission capabilities, and simple interactive features.

<u>References</u>

(1)   Goyal, S. K., and R. W. Worrest, "Expert System Applications to Network Management," in *Expert System Applications to Telecommunications*, Liebowitz, J.(ed.), John Wiley, New York, 1988, p. 1.

(2)   Radecki, S., *Multimedia and Quicktime*, AP Professional, New York, 1994, pp. 83-175.

(3)   Yager, T., *The Multimedia Production Handbook for the PC, Macintosh, and Amiga*, Academic Press, New York, 1993.

Index

2B1Q code 132

a

A. K. Erlang 246
access time 107
accounting management 27
analog transmission systems (ATS)
        120
animation 47
ANSI Standard TI.403-1989 131
Apple Computer 5
application-specific integrated
        circuits (ASICS) 77
applications shell 112
artificial intelligence (AI) 4
asynchronous data subscriber line
        (ADSL) 9, 23, 120, 371
asynchronous transfer mode (ATM)
        14, 228, 331
audio
        audio compression 194
        Audio Interchange File Format
                (AIFF) 219
        audio quality 239
automated message accounting
        (AMA) system 13, 82
automated teller machine (ATM) 343

b

binary coded decimal (BCD) 218
BISYNC protocol 187
bit mapping 187
bit-mapped images 51
branched network 322
British Robot Association 66

broadband ISDN 120
bus network 322
Butterworth filter 59

c

carrier serving area (CSA) 128
CCITT H.263 27
CCITT H.322 27
CD-ROM 338
charge coupled diodes (CCDs) 47
circuit-switched network 318
coding
        algorithms 185
        interframe coding 172
        intraframe coding 172
        linear predictive coding (LPC)
                175
color printing 377
Common Management Information
        Service (CMIS) 275
Common Management Interface
        Protocol (CMIP) 273
common applications service element
        (CASE) 306
common interface format (CIF) 28
community-access TV system (CATV)
        23, 123, 138
compression 21
        compression ratio 187
        diatomic encoding 191
        Huffman coding 183
        lossless algorithms 170
        null compression 187
        run-length coding 182
computer applications
        computer-aided design (CAD)
                66
        computer-aided instruction
                (CAI) 67

computer-assisted layout and composition 379
computer-integrated manufacturing (CIM) 66
computer graphics adapter (CGA) 7
conditioning 133
configuration management 27
continuous quality assessment 282
cost-effective binding system 377
customer premises equipment (CPE) 137
customer proprietary network information (CPNI) 368

d

data communications equipment (DCE) 295
data compression 186, 380
data storage
  buffers 110
  cylinders 48, 98
  data blocks 48, 99
  data count 103
  key field 100
  interblock gap 102
  intrablock gap 103
  tracks 98
database management
  data dictionary 112
  data loading 127
data fusion techniques 366
data terminal equipment (DTE) 295
  database 115
deadlocks 295
decentralized network management 147
decision processes
  decision space 255
  decision tree 253
deductive reasoning 257
Defense Advanced Research Projects Agency (DARPA) 303
Department of Defense (DoD) 273
differential pulse code modulation (DPCM) 169, 174
digital technology
  access cross-connect switch (DACS) 355

facsimile 193
  signal processors (DSPs) 6
  technology requirements 321
frame displacement
  displaced-frame difference 172
  displaced-frame distance (DFD) 174
distributed, intelligent, multimedia, object-oriented DBMS 378
document scanners 50
domain-specific information 65

e

effective information transfer ratio (EITR) 186
encapsulated post-script (EPS) 218
error correction 127
error reporting and evaluation 232
expert systems 64
exponential filter 59
extended binary coded decimal (EBCDIC) 218
extended graphics adapter (EGA) 7

f

fault management 26
FAX 193
FDDI 35, 125
FDDI-II 125
fiber optic systems 76
fiber optics 76
figure of merit 187
File Transfer Access Management (FTAM) 159, 306
firmware 295
flat panel display 77
fourier transform 90
fractional T-1 351, 371
frame relay 35, 37, 126

h

high-speed data subscriber line (HDSL) 23, 120, 371
HDTV 2
high-speed printing 376

horizontal blanking 78
human interface 113
Hypercard 8
hypermedia 4
hypertext 4

i

IBM's Netview 147
IEEE 802.2 125
image
    description 23, 122
    equalization 55
    processing 379
    smoothing 55
image formats
    graphics interchange format
        (GIF) 28, 217
    picture format (PICT) 28, 218
    tagged image file format
        (TIFF) 28
IMS 218
in-band signaling 349
inductive reasoning 258
information services
    America On-Line 2, 13
    Compuserve 2, 13
    Prodigy 2, 13
information superhighway 2
Integrated services digital network
    (ISDN) 137
interactive video 74
interactive voice response (IVR) 68
interface activity 294
International Standards Organization
    (ISO) 52, 174, 272
Internet 334
Internetwork Packet Exchange (IPX)
    154
interoperability 114
    bridges 289
    gateways 289, 290
    routers 289
isochronous data flow 16

j

Joint Picture Experts Group (JPEG)
    9, 188, 196

k

kiosk 338

l

local area network (LAN) 322
language interface 113
LANTERN 159
layer management entities (LMEs)
    275
light-emitting diode (LED) 50
line replaceable units (LRUs) 262
link access protocol (LAP) 297
livelock 295

m

Macintosh File System (MFS) 157
management information base (MIB)
    305
Manufacturing Automation Protocol
    (MAP) 306
mean squared error (MSE) 174
mean time between failures (MTBF)
    260
medium access control (MAC) 123
message-switched network 317
message-oriented protocols 295
metropolitan area networks (MANs)
    35, 76, 159
Microsoft 5
Microsoft Word 7
MIDI 8, 238
mixed media 4
Motion Picture Experts Group
    (MPEG) 28, 77, 174
motion estimation 174
motion vector 174
movies-on-demand 354
multimedia
    applications 282
    multimedia PC (MPC) 6, 22
    network 282
    "phone booth" 354
    platforms 206
    servers 204

multimode fiber 93
multiplexing 36
MYCIN 66

n

Narrow-band transmission systems
        47
Netware LANalyzer 161
Netware Loadable Modules (NLMs)
        157
Network Client Extension (NETX)
        157
Network File System (NFS) 159
Network Management Command
        and Control System (NMCC)
        159
network management systems 273
network operating system (NOS) 157
networks 204
North American Television Standards
        Committee (NTSC) 78
Novell Netware 161
NYNEX 13
Nyquist criterion 195

o

OC-3 (fiber optic transmission
        standard) 121
on-line information systems 338
Open Systems Interconnect (OSI)
open systems interconnect (OSI)
        model 272
optical character recognition (OCR)
        50
Oracle 219
out-of-band signaling 349
overhead 36

p

packet assemblers and dissemblers
        (PADs) 228
packet switching 15
packet-oriented protocols 295
packet-switched network 317
PAL 87

pay-per-view (PPV) 351
pattern substitution 192
performance management 27
personal communications systems 13
personality profile generation 380
pixel 188
platter 99
points of presence (POPs) 147
communication protocols 290, 295
Poisson distribution 249
polynomial regression 244
procedures 255
production systems 256
PROLOG 256
protocols
        converters 297
        Universal detection and
                conversion 282
pulse code modulation (PCM) 93,
        164

q

quadrature amplitude modulation
        (QAM) 132
quadrature phase shift key (QPSK)
        47
quarter common interface format
        (QCIF) 28
Queuing theory 350
Quicktime 8

r

relative encoding 192
repeaters 365
ring network 321
RISC (reduced instruction set
        computer) 267
Robot Institute of America 66
RS-232 288
RS-449 288
RS-488 294

s

sample and hold 167
scanning/OCR 378

SCSI communications channel 7
SECAM 87
sectors 99
security network management 27
semantic net 253
sequential file 101
service advertising protocols (SAP)
        162
signal averaging 60
Simple Gateway Management
        Protocol (SGMP) 303
Simplified Network Management
        Protocol (SNMP) 157, 273
single-mode fiber 93
slow-scan TV 26
SMDS 35
SONET 23, 331
specific application service element
        (SASE) 306
SQL/DS 219
SRI 66
star network 319
state transition graph 254
station message detail record
        (SMDR) 82
storage 110
system management application
        entity (SMAE) 275
Systems Network Architecture (SNA)
        149

t

T-1 service 130
Tagged Image File Format (TIFF) 216
TCP/IP 126, 127, 153
technologies 282
text quality 241
The Wall Street Journal 13
Timbuktu 34
time-division multiplexer 191
transaction control support 381
transmission leg 125
transmission rate 107
transmission system 205
trapezoidal filter 59

u

Unified Network Management
        Architecture (UNMA) 144
user network interface (UNI) 122

v

vertical blanking 78
very-high-scale integrated circuit
        (VHSIC) 77
Very-large-scale integrated (VLSI) 77
VGA 242
video
        Video for Windows 8
        Video quality 236
        video dial tone 82
        video frame 254
        video graphics adapter (VGA)
                7
        video jukebox 46
        video playback 110
virtual circuit 156
VISCA 8

w

wide area networks (WANs) 35, 76,
        159
wide-bandwidth transmission
        systems 47
wireless technology 76
write once-read many (WORM) 45

x

X.25 44, 127
X.400 300
X.500 300

ABOUT THE AUTHOR

Larry L. Ball is currently managing director of Telenetic Controls in Vancouver, British Columbia, Canada. He has more than 20 years of experience in the management of large-scale system development and high-technology product integration programs in such industries as aerospace, medical electronics, and consumer telecommunications products. His technical areas of expertise include satellite communications, computer simulations, digital messaging systems, and intelligent systems. Dr. Ball is also the author of *Cost-Efficient Network Management* and *Network Management with Smart Systems*, both published by McGraw-Hill.